AF595334

Development of Sustainable Energy

Development of Sustainable Energy: Generation Technologies and Concepts

Editors

Mehreen Saleem Gul
Eulalia Jadraque Gago
Tariq Muneer

MDPI • Basel • Beijing • Wuhan • Barcelona • Belgrade • Manchester • Tokyo • Cluj • Tianjin

Editors
Mehreen Saleem Gul
Heriot-Watt University
UK

Eulalia Jadraque Gago
Universidad de Granada
Spain

Tariq Muneer
Edinburgh Napier University
UK

Editorial Office
MDPI
St. Alban-Anlage 66
4052 Basel, Switzerland

This is a reprint of articles from the Special Issue published online in the open access journal *Energies* (ISSN 1996-1073) (available at: https://www.mdpi.com/journal/energies/special_issues/generation_technologies_and_concepts).

For citation purposes, cite each article independently as indicated on the article page online and as indicated below:

LastName, A.A.; LastName, B.B.; LastName, C.C. Article Title. *Journal Name* **Year**, *Article Number*, Page Range.

ISBN 978-3-03943-320-9 (Hbk)
ISBN 978-3-03943-321-6 (PDF)

Contents

About the Editors

Mehreen Saleem Gul is an Assistant Professor in Architectural Engineering in the School of Energy, Geoscience, Infrastructure and Society at Heriot-Watt University, Edinburgh. Mehreen's research experience is in environmental, engineering, and social issues associated with renewable and low-carbon technologies and sustainable buildings. She has developed her own outdoor research labs to investigate the performance of various photovoltaics including bifacial and the impact of ground albedo. Mehreen has over 34 peer-reviewed articles and is a contributor to the Chartered Institution of Building Services Engineers (CIBSE): Guide J, Weather, Solar and Illuminance data (2002) and Guide A, Environmental Design (2015). Mehreen was awarded the CIBSE Napier Shaw Bronze Medal for best paper on an entirely new approach for estimating solar diffuse irradiance.

Eulalia Jadraque Gago is a Professor at the University of Granada, in the Department of Civil Engineering Construction and Engineering Projects. She holds a Civil Engineering, MEng and PhD and has contributed to 23 international publications, and 4 chapters in books of prestigious publishers, such as CIBSE Guide A: Environmental design. Eulalia has over 15 international conferences and provided continued participation in investigation projects and contracts acting as a researcher. Eulalia has directed 1 thesis and has had management charges at the University. Eulalia has been an invited researcher and professor at different universities. Since 2013, she has been a member of the World Society of Sustainable Energy Technologies

Tariq Muneer (Professor) is a Professor of Energy Engineering at Napier University, Edinburgh and currently chairs an active group engaged in research on 'Sustainable Energy' that includes 'Sustainable Transport'. Professor Muneer is an international authority on the subject of solar energy and its use in buildings with over 35 years of experience. He is the author of over 215 technical articles, most of which have been distilled in his research monographs.

Preface to "Development of Sustainable Energy: Generation Technologies and Concepts"

The lack of access to energy supplies and transformation systems is a constraint to human and economic development. Achieving solutions to environmental problems that we face requires long-term potential actions for sustainable development. Sustainable energy generation should be widely encouraged, as it does not cause any harm to the environment and is widely available free of cost. Harnessing renewable energy resources appears to be one of the most efficient and effective solutions. Renewable energy sources such as solar, wind, geothermal, hydropower, biomass, and marine energy are at the centre of the transition to less carbon-intensive and more sustainable energy systems. The use of renewable energy through improved technologies and concepts is, therefore, expected to play a major role in the future of sustainable energy generation. Energy is central to nearly every major challenge and opportunity the world faces today. Be it for jobs, security, climate change, food production, or increasing incomes, access to energy for all is essential. The adoption of the new United Nations Sustainable Development Goals (SDGs) in 2015 marked a new level of political recognition of the importance of energy for development. For the first time, this included a target to ensure access to affordable, reliable, sustainable, and modern energy for all—collectively known as Sustainable Development Goal 7, or SDG 7. The aim is to enhance international cooperation by 2030 to facilitate access to clean energy research and technology, including renewable energy, energy efficiency, and advanced and cleaner fossil-fuel technology, and to promote investment in energy infrastructure and clean energy technology. The challenge is, however, far from being solved, and there needs to be more access to clean fuel and technology. Furthermore, more progress needs to be made regarding the integration of sustainable energy into end-use applications in sectors such as building, transport, and industry. This edition of the *Energies* journal addresses the barriers and challenges facing sustainable energy generation for future energy technologies and concepts and highlights potential solutions that should lead to sustainable development.

Mehreen Saleem Gul, Eulalia Jadraque Gago, Tariq Muneer
Editors

Article

The Role of Low-Load Diesel in Improved Renewable Hosting Capacity within Isolated Power Systems

James Hamilton *, Michael Negnevitsky, Xiaolin Wang and Evgenii Semshchikov

Centre for Renewable Energy and Power Systems, School of Engineering, University of Tasmania, Hobart 7000, Australia; michael.negnevitsky@utas.edu.au (M.N.); xiaolin.wang@utas.edu.au (X.W.); evgenii.semshchikov@utas.edu.au (E.S.)

* Correspondence: james.hamilton@utas.edu.au; Tel.: +61-3-6226-2685

Received: 22 June 2020; Accepted: 28 July 2020; Published: 5 August 2020

Abstract: Isolated communities are progressively integrating renewable generation to reduce the societal, economic and ecological cost of diesel generation. Unfortunately, as renewable penetration and load variability increase, systems require greater diesel generation reserves, constraining renewable utilisation. Improved diesel generator flexibility can reduce the requirement for diesel reserves, allowing increased renewable hosting. Regrettably, it is uncommon for utilities to modify diesel generator control during the integration of renewable source generation. Identifying diesel generator flexibility and co-ordination as an essential component to optimising system hosting capacity, this paper investigates improved diesel generator flexibility and coordination via low-load diesel application. Case study comparisons for both high- and low-penetration hybrid diesel power systems are presented in King Island, Australia, and Moloka'i, Hawai'i, respectively. For King Island, the approach details a 50% reduction in storage requirement, while for Moloka'i the application supports a 27% increase in renewable hosting capacity.

Keywords: battery storage; hybrid power system; low-load diesel; microgrid; remote area

1. Introduction

Isolated power systems (IPSs) have historically relied on diesel generation given the accessibility, reliability and maintainability of the technology. More recently IPSs have started to integrate renewable generation, as awareness of the economic and environmental impacts of diesel generation have become known [1,2]. Wind and solar photovoltaic (PV) represent the two most common renewable technologies employed to reduce diesel consumption, however, both are stochastic, and unable to eliminate diesel generation entirely [3,4]. To eliminate diesel generation, enabling technologies such as energy storage are required. Unfortunately, storage is currently expensive and complex, making it unsuited for the majority of IPSs [5,6]. In response, a number of IPSs with improved generation and load control have been created to mitigate the need for storage [7–9]. To this end, this paper assesses the role of low-load diesel within two innovative case studies, King Island, Tasmania, a high-penetration wind IPS, and the island of Moloka'i, Hawai'i, a low-penetration solar PV IPS with ambitious near-term renewable targets. The novelty of this paper lies in identifying alternative approaches to energy storage integration, validating this approach via case study review within both wind- and solar-dominated IPSs. The case studies selected, Moloka'i Hawai'I, and King Island, Tasmania, Australia, represent the current best practice for renewable integration, Figure 1.

Owing to their small size, high diesel fuel cost and resilient communities, islanded IPSs are some of the earliest adopters of renewable generation technologies [10]. King Island is a case in point, representing one of the world's first megawatt-scale highly renewable penetration power systems. Early adopters, such as King Island, have an important role to play in the adoption, testing and commercialisation of renewable generation and enabling technologies. The challenge in leveraging and redeploying experiences, such as those obtained on King Island, has always been how to scale these approaches for larger markets while consolidating capital cost [11].

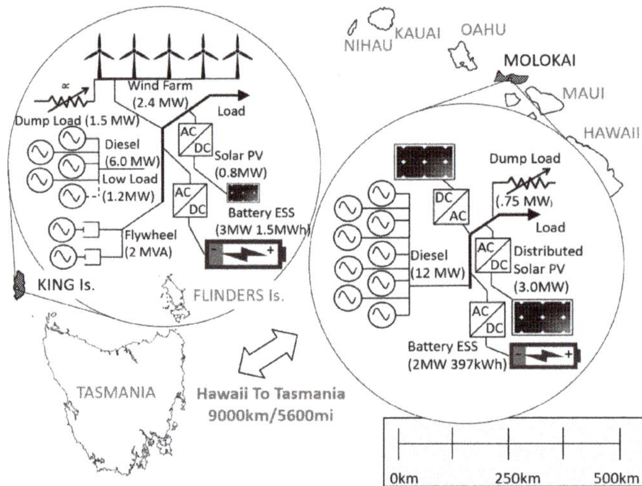

Figure 1. King Island isolated power system schematic, left hashed, and Moloka'i isolated power system schematic, right hashed.

Larger systems are generally slower to reach high renewable penetrations, owing to the large capacity of renewable generation required. In this regard, the lessons learnt from IPSs can both accelerate and derisk renewable integration in larger markets. The island of Moloka'i, Hawai'i represents one such case study. Larger than King Island, and reliant on residential solar PV instead of wind, the island is looking to scale existing high renewable penetration experience, leveraging technologies such as those deployed on King Island to meet local targets for 100% renewable generation. In reviewing the technology options available to Moloka'i, this paper presents a general introduction to the legacy technology progression within Hawai'i and Australia in Section 1.1, ahead of a case study review in Sections 1.2 and 1.3. The paper's modelling methodology, results, and conclusions are presented in Sections 2–4, respectively.

1.1. The Technology Legacy of Hawaiian and Australian Isolated Power Systems.

Wind was first pioneered in Hawai'i as part of the US Department of Energy's federal wind program. Administered by NASA, the program targeted the realisation of a sub 5 c/kWh levelized energy cost. This pioneering experience was quickly followed by multiple wind turbine developments across the islands of O'ahu, Maui, Hawai'i and Moloka'i, Table 1. For most of the 1980s, Hawai'i led global wind technology development, hosting the world's largest wind turbine, the Boeing Mod-5B, a 3.2 MW twin blade turbine presenting an impressive 97 m diameter rotor. The heyday of Hawaiian wind development, unfortunately, came to an end shortly afterwards, signaling a loss in social license for wind development across the islands [12]. In stark contrast, Hawai'i has enthusiastically embraced solar PV, supported by attractive resource and net energy metering policy (2001–2015). Perhaps most importantly for Hawai'i, solar PV has also proven to be highly modular and scalable, placing the technology within community reach. Hawai'i currently generates approximately 11.2% of its load via combined centralised and distributed solar PV schemes [12]. This compares favourably to Australian and US averages of 5.2% and 2.3%, respectively [13,14]. The uptake has been so successful that, for many of the Hawaiian Islands, the adoption of solar PV has reached, or is rapidly approaching, the system's hosting capacity. Moloka'i is a case in point, exhibiting instantaneous midday solar penetrations exceeding 75%, despite a relatively low annual penetration of below 14%.

Table 1. Hawai'i wind power developments 1980–2020.

Name	Commissioned	Decommissioned	Location	Capacity	Turbine
Kahuku	1980	1982	O'ahu	200 kW	MOD-0A 200 kW
Kahua	1983	1992	Hawai'i	3.4 MW	Jacobs 17.5 kW
Windane	1984	1991	Maui	340 kW	Windane-31
Kahuku	1985	1996	O'ahu	9 MW	Westinghouse 600 kW
Lalamilo	1985	2010	Hawai'i	2.3 MW	Jacobs 17.5/20 kW
Kahuku	1987	1993	O'ahu	3.2 MW	MOD-5A 3200 kW
Kama'oa	1987	2006	Hawai'i	9 MW	Mitsubishi 250 kW
Moloka'i	1991	1997	Moloka'i	300 kW	Vestas V20
Hawi	2006	ongoing	Hawai'i	10.56 MW	Vestas V47 600 kW
Kaheawa	2006	ongoing	Maui	30 MW	GE 1.5 MW
Pakini	2007	ongoing	Hawai'i	20.5 MW	GE 1.5 MW
Pakini Nui	2007	ongoing	Hawai'i	20.5 MW	GE 1.5 MW
Kahuku	2011	ongoing	O'ahu	30 MW	Clipper 2.5 MW
Kaheawa II	2012	ongoing	Maui	21 MW	GE 1.5 MW
Auwahi	2012	ongoing	Maui	21 MW	Siemens 2.1 MW
Kawailoa	2012	ongoing	O'ahu	69 MW	Siemens 2.3 MW

Effective in displacing diesel generation, however in general, their performance offered poor reliability and reliance [15]. While early Australian trials were not to the extent of the Hawaiian research, they ushered in Australia's first wind farm at Salmon Beach, Western Australia (1987), and then a second at Huxley Hill, King Island, Tasmania (1998). In contrast to the Hawaiian experience, Australia's slow uptake found roots, with wind technology transitioning to broader network application. Tasmania currently generates approximately 9.7% of its total load via wind generation, ahead of the Australian and US national averages of 7.1% and 6.5%, respectively [13,14]. In contrast, 4.9% of Hawai'i's energy is wind derived, despite modern wind projects realising the 5 c/kWh price point originally envisaged by the US department of energy [16]. For reference, Hawai'i's energy tariffs currently exceed 28 c/kWh across most islands.

Australia also adopted wind turbine technology during the 1980s, wind representing the only viable renewable technology of the era. The best of these early wind turbines proved to be cost

In this regard, our two case studies (Table 2) are representative of broader regional preferences, with Moloka'i a solar-dominated decentralised IPS and King Island a wind-dominated centralised IPS.

Table 2. Island power systems case study metrics.

	King Island Renewable Energy Integration Project	Moloka'i Secure Renewable Microgrid
Developer/Owner	Hydro Tasmania	Maui Electric Company
Peak Load (MW)	2.5	5
Average Load (MW)	1.4	3.7
Annual Generation (GWh p.a.)	12	32
Population	1600	8000
Annual Tourist Numbers p.a.	7000	80,000
Distance to major port (km)	250	40
Diesel Capacity (MW)	6	13
Wind Capacity (MW)	2.25	0
Solar PV Capacity (MW)	0.8	3
BESS Capacity (MW/MWh)	3/1.5	1/0.397
Flywheel System	Yes	No
Renewable Energy Penetration (% p.a.)	65%	13%
Development Period	1998–2015	2009–ongoing
Utility inter-island connection (cable)	No	No

1.2. King Island

Hydro Tasmania initially hybridised the King Island IPS to explore wind technologies in the late 1990s, with the system subsequently expanded via the integration of solar PV, batteries and flywheel technologies. This section considers the King Island experience, in particular efforts to reduce the cost and complexity of many of the developed applications.

Situated between Victoria and Tasmania, Australia, King Island is located within the strategic shipping channel of Bass Strait. The wind resource on King Island, averaging 9.0 m/s at 60 m elevation, is now employed to power King Island's IPS. Over 50% of King Island's annual demand is met via the 2.45 MW of installed wind capacity. The remainder of King Island generation consists of 0.8 MW of uncontrolled residential solar PV and 7.2 MW of diesel generation, Table 3 [17]. Annual renewable generation, wind and solar combined, contributed over 60% of the islands load last year, Figure 2, with the system running diesel off for 20% of the year. To support system security a range of enabling technologies are employed on King Island, including a 3 MW, 1.5 MWh advanced lead acid battery energy storage system (BESS), a 2 MVA diesel coupled flywheel energy storage system, 0.1 MW of residential demand side management and a 1.5 MW resistive dump load. The King Island IPS is managed by an automated IPS controller, allowing the system to operate unattended.

Table 3. King Island Currie Power Station generation.

Unit	Generator	Governor	MW	RPM	Cylinders	Age (Yrs)
G01	Caterpillar 3516B	CAT ADEM, Woodward AGLC	1.6	1500	16	12
G02	Caterpillar 3516B	CAT ADEM, Woodward AGLC	1.6	1500	16	12
G03	Caterpillar 3516B	CAT ADEM, Woodward AGLC	1.6	1500	16	22
G04	Caterpillar 3516B	CAT ADEM, Woodward AGLC	1.2	1500	16	34
G05	MTU S4000	ComAp InteliSys	1.2	1500	16	2

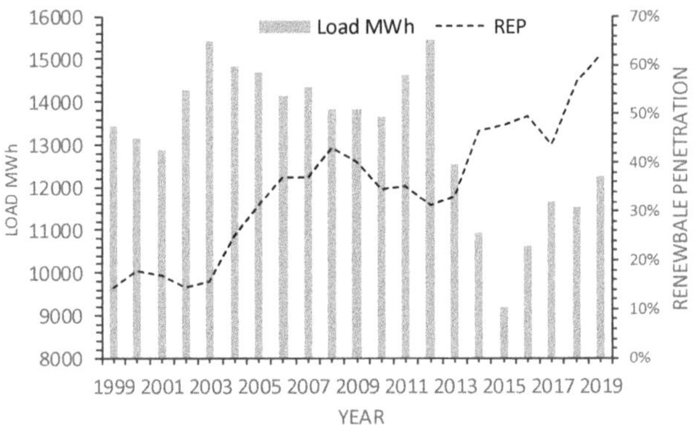

Figure 2. King Island isolated power system performance from 1999 to 2019.

The dispatch strategy adopted on King Island targets the maximum utilisation of the available wind generation. Solar PV generation is not a large determinant within the control methodology, as the utility has no visibility or control of this component of the system. Instead, dispatchable generation is scheduled to respond to load and resource variability. The diesel dispatch strategy progressively adds diesel capacity interchangeably, with the exception that the first diesel on and the last diesel off is the MTU low-load unit. In contrast the CAT engines, which adopt a 40% low-load limit, the MTU is warranted to 10% loading, assisting renewable penetration under high wind contribution.

Huxley Hill wind farm was commissioned in 1998, initially consisting of three Nordex N29 wind turbines (0.75 MW total). Huxley Hill wind farm initially reduced diesel consumption by one fifth. Encouraged, Hydro Tasmania integrated additional renewable capacity in 2004, adding two Vestas V52 turbines (1.7 MW total). In support, a 200 kW, 800 kWh vanadium redox flow battery (VRB) was also integrated the same year. The VRB uses aqueous vanadium electrolytes separated by a proton exchange membrane. Ion exchange provides an energy storage concept offering long service life and tolerance to high cycle rates and depths of discharge. Unfortunately, the flow battery proved complex and difficult to maintain. The failure of the VRB electrolyte containment resulted in the decommissioning of the battery shortly after installation. Without storage, system operation required the set point control of the wind production (renewable spillage) to ensure system security. Despite the failure of the VRB, renewable penetration exceeded one third of the system load, with King Island able to demonstrate medium levels of renewable penetration from 2005.

This milestone signified an important achievement for renewable integration within Australia. Leveraging this experience, Hydro Tasmania then embarked on a period of research and development encompassed by the King Island renewable integration program. King Island allowed Hydro Tasmania to assess a range of emergent renewable technologies, including, solar photovoltaics (2008), concentrated solar thermal (2009), flywheel energy storage (2011), biodiesel (2012), residential load shedding (2012), battery storage (2014), low-load diesel (2015) and wave generation (2020). The King Island renewable integration program is primarily responsible for the current system performance, exceeding 60% penetration. The technology successes and failures observed across this period remain relevant to a range of current applications and markets, including both isolated and networked power systems globally. Of the technologies to fail, both the concentrated solar thermal and dual axis solar PV systems were early casualties. The solar PV tracking failed due to repeat failures within the hydraulic tracking mechanism, and despite the solar PV panels being unaffected, remediation costs have prevented system reinstatement. In contrast, the concentrated solar thermal project, envisaged to consist of six 19T elevated graphite solar storage receivers, was never implemented. Ironically, the dynamic resistive heating element, developed as a complementary heating source has evolved into a flexible enabling technology in its own right. During the testing of the resistive load, it became clear that the fast and accurate response provided value as a dispatchable load, specifically offering fast frequency raise reserve. The adoption of a resistive dump load allowed Hydro Tasmania to operate wind generation unconstrained, providing improved system inertia and capacity firming. The resistive dump load provides capacity firming via either dispatch or the withdrawal of the load in a fraction of a second, as required to smooth the accompanying wind generation. In this manner, the resistive load contributes to frequency regulation, lowering the reserve requirements, and reducing diesel consumption and maintenance. The role of the resistor is illustrated in Figure 3, where control transfers between the battery, resistor and diesel generation. The plot shows a transition from battery charging (hour 1), to discharging (hour 3). The system load dips mid-plot, during which the resistive load is used to spill surplus renewable generation. As the load increases, the battery resumes control, injecting energy until a drop in wind production triggers a diesel start (hour 4). Prior to this, the system was running diesel off.

The transient response of the BESS and resistive load is further illustrated in Figure 4, inclusive of flywheel energy exchange. The plot covers a two-minute interval of steady system load, approximating 1.7 MW. A rapid drop in wind generation (t = 20 s) requires inertia support from the flywheel and a discontinuation of resistive spill. The battery responds to inject energy shortly after, allowing the flywheel to resume 50 Hz operation. Had the battery state of the charge been insufficient, the flywheel would have coupled to a paired diesel engine, performing a diesel fast start via the integrated mechanical clutch. In this instance the combination of the battery, resistor and flywheel mitigated the need to bring diesel generation online. The flywheel technology consists of two 12 T horizontal steel flywheels, each coupled to a 1 MVA diesel generator. These engines are not used outside of the provision of fast start diesel response, given the large mechanical loads and reduced service life imposed during the engagement of the mechanical clutch. The role of the flywheel within the King Island IPS is to provide inertia and a fast-start diesel contingency to both dampen and respond to variable renewable output. The fast start diesel response occurs over a few seconds, with the load transferred to the conventional diesel assets as they are brought online over a matter of minutes.

The coupling of the flywheel to a diesel engine provides a system response extending beyond the 30 s of inertia available from the two flywheels. During diesel off operation, both flywheels will be operational. At other times of high renewable penetration, a single flywheel will typically be operational. Under medium or low renewable penetration, the flywheels are turned off to reduce the ~60 kW of parasitic load required to keep each unit spinning.

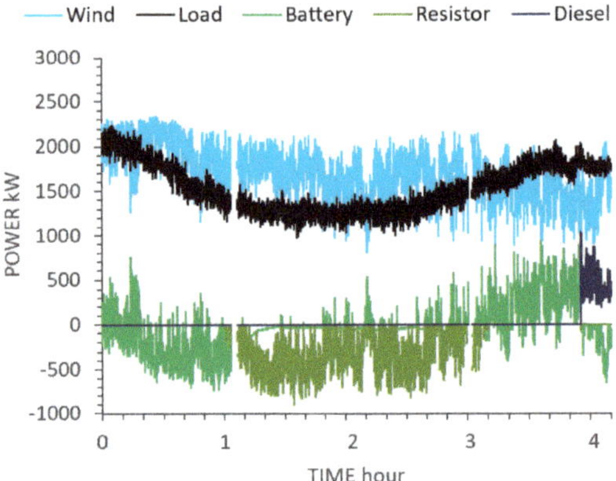

Figure 3. King Island generation showing system transition between wind (light blue), battery (light green), resistor (dark green, 1–3 h) and diesel (dark blue, post 4 h) isochronous control.

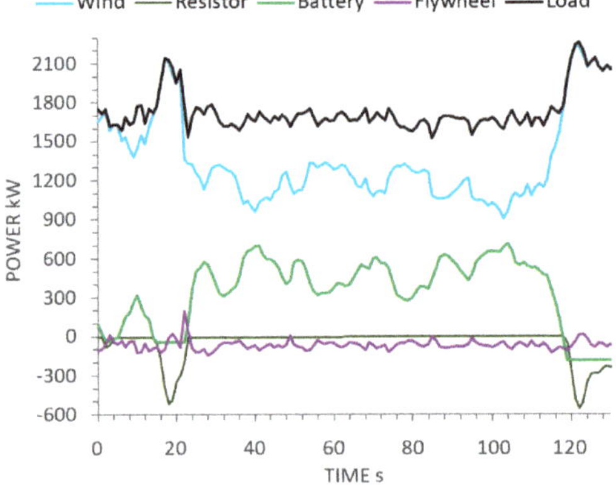

Figure 4. King Island flywheel (purple) response to rapid loss of wind generation (light blue), supporting battery response (light green) and ride through.

To allow for extended diesel off operation, a 3 MW, 1.5 MWh advanced lead acid BESS was integrated in 2014. The battery extended the time for which the system could remain diesel off, having accumulated over 12,000 h of diesel off operation to date. A typical daily load and generation profile is shown in Figure 5, with the twin peak load profile evident. The plot shows a five-hour diesel off period from midday, as the afternoon sea breeze produces surplus renewable generation. Throughout this period, the battery transitions from an energy source (hours 9–10) to a sink (hours 10–11), until the resistor is deployed to spill surplus renewable (hours 11–1). As the afternoon load peaks, diesel generation is again brought back online. The battery charges into the evening, indicating the inability of diesel generation to reduce its output below its operational low-load limit. The role of low-load diesel within the King Island power system is to reduce the occurrence of excess renewable generation. This is achieved by permitting the low-load diesel unit to run down to 10% of its rated capacity during periods of high wind generation.

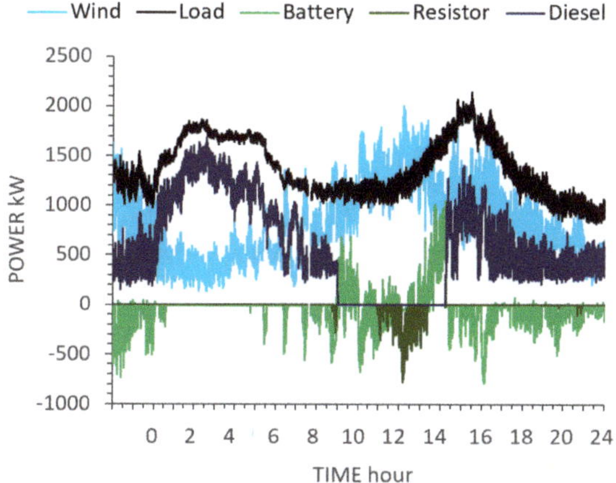

Figure 5. Twenty-four hours of King Island generation, showing diesel off operation as the afternoon sea breeze comes on.

1.3. Moloka'i Hawai'i

Maui Electric Company (MECO) have established a target for Moloka'i to achieve 100% renewable energy (RE) by 2020 [18]. In doing so, Moloka'i will be the first Hawai'i island to reach this milestone, setting a roadmap for the other islands of Hawai'i and US states to follow. This section considers to what role coordinated generation can be utilised in support of these goals. Modified diesel application is assessed via simulation of a low-load diesel operating scenario. This paper addresses operational dependencies between generation, load and storage, quantifying to what extent generation and load flexibility can provide improved near-term renewable hosting capacity.

The island of Moloka'i is located centrally within the Hawaiian archipelago, between the larger islands of O'ahu, to the west, and Maui, to the East. Approximately 90 km east of Honolulu, the coastal proximity to both O'ahu and Maui is under 40 km. The island's population is approximately 8000, of which around 40% assert native Hawaiian ancestry. Tourism, cattle, and diversified agriculture represent the island's major economies. The electrical demand on Moloka'i peaks at around 4500 kW, presenting the typical twin peak profile common to many island communities, Figure 6. Notably, in recent years, the increase in the capacity of residential solar PV has had a notably depressed midday load. It is also interesting to note the absence, for the time being, of centralised utility renewable development on the island, in part due to strong local opposition to large-scale development, viewed as incompatible with local customs and culture [12].

Blessed with abundant wind and solar resources, it is somewhat surprising to note the absence of centralised renewable generation on Moloka'i. The scenario is even more surprising considering MECO's early exposure to renewable generation. Unfortunately, the failure of these early projects to be inclusive of community concerns, combined with disinterest from MECO to own and operate renewable assets, have reduced the investment in centralised generation. In contrast, since 2009, the installed capacity of uncontrolled (unable to receive a utility set point) residential solar PV has increased markedly on Moloka'i, so much so that from 2015 to 2018 further solar PV interconnection was restricted, with the system's hosting capacity for uncontrolled solar PV saturated. Solar PV hosting capacity is the ability of the system to accept additional PV generation without pushing the midday load below the systems reserve requirements, as set by the minimum load setpoint of the systems thermal generation, P_{min}. In response, MECO in partnership with the Hawai'i Natural Energy Institute (HNEI) installed a number of enabling or ancillary technologies, including, a 2 MW 397 kWh lithium

BESS and a 750-kW resistive load bank. These technologies were instrumental in relief to hosting capacity constraint, allowing the solar PV interconnection queue to reopen. Both technologies allowed for a reduced reserve requirement, providing system flexibility to manage solar resource variability. The impact of additional solar PV capacity is evident in Figure 6, where a Moloka'i daily load profile is shown inclusive of the solar PV as negative load. Notably, the reduced midday minimum loading evident in 2018, approaches P_{min}, suggesting a limited ability to further integrate additional solar PV capacity. Despite midday instantaneous solar PV saturation, annual average solar PV penetration remains low at around 14%. Considering the aggressive development timeframe outlined for a 100% renewable transition, MECO are prioritising integration of near-term enabling technologies, in parallel with discontinued replacement or purchase of diesel generation. The generators supplying Moloka'i's Pālā'au Power Station, Table 4, consist of a range of high-, medium- and low-speed diesel generation.

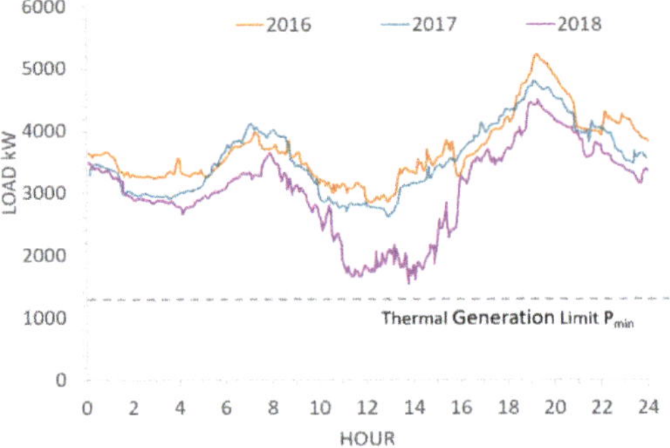

Figure 6. Moloka'i average daily load profile 2016 (orange), 2017 (blue) and 2018 (purple). Note the reduction in midday load under increasing solar photovoltaic (PV) penetration [19].

Table 4. Moloka'i's Pālā'au Power Station generation.

Unit	Generator	Governor	MW	RPM	Cylinders	Age (Yrs)
G01	Caterpillar 3516	Woodward 2301A	1.25	1800	16	34
G02	Caterpillar 3516	Woodward 2301A	1.25	1800	16	34
G03	Cummins KTA50	American Bosch CU673C-17 A	0.97	1200	16	34
G04	Cummins KTA50	American Bosch CU673C-17 A	0.97	1200	16	34
G05	Cummins KTA50	American Bosch CU673C-17 A	0.97	1200	16	28
G06	Cummins KTA50	American Bosch CU673C-17 A	0.97	1200	16	28
G07	Caterpillar 3608	Woodward 2301D	2.2	900	8	23
G08	Caterpillar 3608	Woodward 2301D	2.2	900	8	23
G09	Caterpillar 3608	Woodward 2301D	2.2	900	8	23

In general, high speed engines typically offer an improved generator response, while lower speeds offer improved inertia and peak efficiency. Heat release curves for all units were provided by the MECO, with efficiency varying significantly between engines. Part of this variation can be attributed to the significant age of the diesel asset base. Where outlier unit efficiencies were identified this data was flagged for low reliance and not used within subsequent analysis. Moloka'i's diesel asset base shares similarities with that of King Island, with a number of the King Island strategies adopted in advancing Moloka'i's renewable transition.

The role of the Moloka'i battery is to provide fast-acting coordinated frequency support, improving system stability by providing the diesel generators time to respond. In this manner, the battery state of charge is maintained at 50% to provide for both the frequency rise and the lower reserve. The battery is provided with six raised and six lower set points, defining a rapid step response followed by a gradual load transfer in both over and under frequency events, Figure 7. Significant effort has been undertaken by HNEI to improve the response time of the BESS, with a revised control architecture reducing the response latency from ~250 ms to ~60 ms, [20]. The Moloka'i battery is the third battery ESS installed in partnership between HECO and HNEI, with batteries also in place on O'ahu and Hawai'i island [21]. All three batteries target fast response frequency regulation or power-smoothing applications. Additional issues encountered across BESS integration include, high inverter temperatures and communications faults. Inverter temperatures were resolved by constraint of the inverter's reactive setpoint. Communication issues were resolved via hardware replacement.

The role of the load bank is to manage the system frequency during periods of excess solar PV generation, subject to BESS state of charge. In this application, the load bank reduces Moloka'i's reserve requirement, providing a discretionary load to balance the grid during periods of excess energy. In assessing what additional near-term applications may further benefit Moloka'i, this paper explores the role of low-load diesel to support generator flexibility and co-ordination.

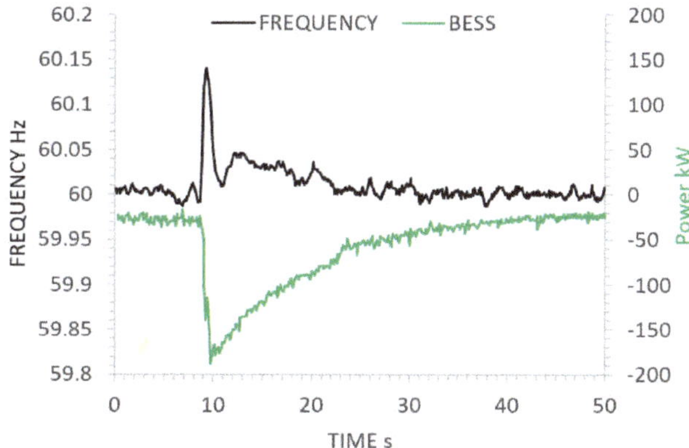

Figure 7. Moloka'i isolated power system (IPS) frequency (black) and battery (green) response to loss of load [20].

2. Methods and Materials

2.1. Modelling

Low-load diesel application was simulated using Homer Pro software, developed by the U.S. national renewable energy agency to assist in the selection and sizing of power generation technologies. The model accepts the daily, seasonal and yearly profiles for resource and load, allowing the user to model diesel and renewable generation via control of generation dispatch order, reserve requirements, generator curtailment and efficiencies [22]. Homer Pro was selected as the appropriate software environment given the prevalence of this format within industry, providing for reduced barriers to utility review and consideration. For both the King Island and Moloka'i case studies measured resource and load data was used to define a 12-month simulation of hourly generation dispatch. Each case study was configured to represent the as-build system configuration, with observed system performance used to validate the model. Model configuration included generation dispatch and reserve definition as implemented for each case study. Model validation consisted of review across both modelled and observed annualised diesel generation run hours, renewable penetration and fuel consumption. For a known solar irradiance profile, the model develops an hourly solar resource estimate [23]. Solar PV power output is then calculated using Equation (1).

$$P_{pv} = f_{pv} Y_{pv} \frac{I_t}{I_s} \quad (1)$$

where f_{pv} is the derating factor, Y_{pv} is the rated capacity of solar PV (kW), I_t is the solar irradiance, and I_s is one kW per square meter. The derating factor can be used to approximate reduced efficiency as may be experienced in specific configurations or environment conditions. For a known wind resource, the model uses a user-defined wind turbine power curve, the relationship between wind speed and power, to calculate wind generation. To facilitate this, an hourly wind resource profile can be estimated from the average wind speed, Weibull shape factor, autocorrelation factor and diurnal pattern, however, for King Island measured hub height hourly wind resource data was used. Postproduction losses applied to both wind and solar generation include electrical losses and unit availability. When the available renewable generation is insufficient to meet the system load, the model may schedule battery or diesel generation according to the maximum and minimum unit loadings. At all other times the system reserve requirements defined the battery and diesel response. The diesel generator fuel curve is assumed to be linear according to Equation (2), and is used to calculate fuel consumption.

$$F = F_{int} Y_{gen} + F_1 P_{gen} \tag{2}$$

where F_{int} is the fuel curve intercept coefficient, F_1 is the fuel curve slope, Y_{gen} is the unit rated capacity (kW) and P_{gen} is the generator output (kW). The units of F depend on the manufacturers preferred measurement units for fuel, typically either litres or gallons per hour. In regard to the battery performance the model uses the nominal voltage, capacity curve, allowable charge range, roundtrip efficiency and cycle life to simulate any battery contribution. The capacity curve details the discharge capacity of the battery versus the discharge current, and is supplied by the manufacturer. The maximum rate of charge or discharge is specified by the kinetic battery model [24]. A detailed overview of the simulation theory is provided in [25].

For each case study the utilities reported annual fuel consumption, dispatch scheme, generator run hours and renewable penetration were used to validate the model configuration ahead of low-load diesel simulation. For Moloka'I, the generation performance and dispatch model was first established using 2009 generation and load data, representative of the system prior to residential solar PV uptake. To this model, annual reported renewable investment was added iteratively to validate the performance under increasing solar PV penetrations. The approach yielded annual diesel fuel consumption estimates within 3% of the observed performance once measured data was corrected in consideration for the high fuel consumption rate of generator 7 (this unit was removed from the model post-calibration, given its inefficient operation and non-standard performance). For King Island much the same methodology and accuracy were employed/observed with the higher annual resource variability (wind compared to solar PV) addressed via validation of the model against operational data using the 5-year moving average (1999–2019). The models were largely insensitive to economic assumptions given the fuel consumption rates were directly compared across simulations to quantify system performance. Irrespective of this, actual incurred fuel costs were adopted to match real inflation. Low-load diesel application was assessed via revision of the diesel low-load limit from 30% to 10%. The efficiency of existing diesel assets under low-load operation was established in prior studies [26], remaining predominantly linear in relationship.

2.2. Low-Load Diesel

Low-load diesel application affords diesel generation improved range and flexibility, permitting system acceptance of additional renewable generation via a reduction in the diesel engine load limit. The low-load limit is set within the primary engine controller on a case by case basis. No hardware of software replacement is required, resulting in a low complexity, low cost and accessible approach applicable to all diesel generators [27,28]. Poor combustion and cylinder condition are responsible for historical restrictions surrounding low-load operation, however, a number of manufactures now warrant low-load applications, reflecting the increased awareness and viability of the practice [29]. Due to poor low-load efficiencies, fuel consumption per kWh of diesel generation increase at low-load, however, given both the increased renewable capture and the low volume of kWh's produced by diesel generation at low-load, the practice has a net positive reduction in fuel usage, as confirmed via system simulation. Net fuel reductions result, despite reduced efficiency, given the acceptance of greater instantaneous renewable penetration. This occurs given the engines ability to further reduce load as renewable generation increases. Conventionally, once an engine hits its 30% load limit any additional renewable generation is spilt from the system. Under low-load application this additional generation is accepted via further diesel load reduction. The major operational concern in reducing an engine's low-load limit is the risk of reverse power acceptance given the reduced ability of the diesel assets to regulate upward renewable variability. This is commonly managed via the inclusion of a dump load to dissipate excess generation as heat.

3. Low-Load Diesel Modelling Results

For high-penetration hybrid diesel systems such as King Island, low-load diesel permits the acceptance of additional renewable generation from any reserves of surplus generation. Simulation of the annual King Island system performance in this manner identified average diesel fuel savings of 6.3% per annum (on a year by year basis savings varied with renewable resource, above average wind generation would result in increased fuel savings, while lower wind generation would decrease observed fuel savings). It can be seen from the results that a reduction in engine low limit serves to lower the systems diesel reserve requirements, reducing the time engines spend operating at their low-load limit. For King Island the application delivers both improve renewable penetration, via a reduction in renewable spillage, and an associated reduction in the requirement for energy storage [30]. For high renewable penetration systems such as King Island, low-load diesel application does not impact the systems renewable hosting capacity, acknowledging the system already hosts a renewable capacity exceeding its maximum load. For high-penetration renewable energy systems such as King Island, low-load diesel is also observed to rationalise the requirement for energy storage In this case the annual average battery utilisation (amount of energy stored, MWh) is reduced by 50%, allowing for a smaller battery capacity and reduced capital cost.

For low-penetration renewable systems, such as Moloka'i, the role of low-load diesel application is less obvious. Primarily, because low-penetration systems rarely spill renewable generation. Simulation of the Moloka'i system identified a range of alternate low-load benefit, including improved hosting capacity and reduced diesel OPEX. Simulation of the pre-2016 (prior to BESS and dump load integration) Moloka'i hosting capacity determined that the limit of uncontrolled solar PV was 2.5 MW. Simulation inclusive of the BESS and dump load, representing the current Moloka'i system, determined the hosting capacity to be 3.0 MW. A low-load diesel scenario was then considered with the diesel generation permitted to run down to 10% loading. A 10% load limit is considered conservative and was selected considering both the age of the assets, and the experience of the neighbouring Kauai Island utility cooperative in obtaining permits to operate at this level. Simulating a low-load limit of 10%, the system hosting capacity was increased to between 3.5 MW and 3.8 MW. Allowing adoption of low-load operation across only the high-speed engines, units 1 and 2, the hosting capacity was increased to 3.5 MW, a 17% increase. For adoption of low-load operation across diesel engines 1, 2 and 9, the hosting capacity was increased to 3.8 MW, a 27% increase. The results define a role for low-load diesel application in near-term relief of hosting capacity constraint. As the Moloka'i system is functionally at its hosting capacity envelope, and does not generally spill generation (less than half a percent of generation is dissipated via the resistive dump load), no immediate increase in renewable penetration was observed under low-load diesel application. Despite this, fuel savings were delivered via low-load diesel operation, achieved via a modified low-load diesel dispatch scheme. The existing Moloka'i dispatch strategy prioritises large low-speed engines (large), generators 7–9, with small high speed (small) generators 1&2, deployed to address load variability. Band allocations 1 through 5 define the generation intensity, with band increase associated with greater diesel capacity, Table 5. Fuel usage correlates to band allocation, with lower bands consuming less fuel. A modified low-load diesel dispatch scheme, Table 6, shifts the balance of generation from band 3 to band 2, via substitution of large low-speed generation for small high-speed generation. The modified low-load diesel dispatch scheme delivers both a reduction in fuel use, and significantly, a reduced maintenance obligation. The reduced maintenance spend results in lower maintenance costs attributed to small inertia and high-speed engines [31].

Table 5. Moloka'i diesel utilisation for a 3 MW solar PV scenario (current dispatch scheme).

Band	Existing Moloka'i Dispatch Strategy (Load Range)	Band Utilisation
1	ONE small and ONE large generator (<1.2 MW)	1.3%
2	**NO small and TWO big generators (<2.3 MW)**	**30.9%**
3	AT LEAST ONE small and TWO large generators (<3.5 MW)	59.1%
4	NO small and THREE Large generators (<4.5 MW)	7.9%
5	AT LEAST ONE small and THREE large generators	0.8%

Table 6. Moloka'i diesel utilisation for a 3 MW solar PV scenario (low-load dispatch scheme).

Band	Proposed Moloka'i Dispatch Strategy (Load Range)	Band Utilisation
1	ONE small and ONE large generator (<1.2 MW)	1.3%
2A	**TWO small and ONE large generator (<2.6 MW)**	**46.3%**
3	AT LEAST ONE small and TWO large generators (<3.5 MW)	43.7%
4	NO small and THREE Large generators (<4.5 MW)	7.9%
5	AT LEAST ONE small and THREE large generators	0.8%

In addition to improved hosting capacity, low-load diesel application within the current Moloka'i system results in a fuel reduction of 1%, and an 8% reduction in maintenance expenditure, Table 7. The combined annual OPEX reduction is $209,469 p.a., representing 2.7% of total annual operational expenditure. Economic modelling was then extended to consider a future 6 MW solar PV capacity scenario, Table 8. While in violation of the system's current hosting capacity, this scenario is useful in exploring the storage requirements required to support additional solar PV deployment. The proposed low-load diesel methodology reduces this requirement by 43%, approximately halving the storage capacity required for grid security under future high-penetration scenarios.

Table 7. Performance of existing and proposed low-load dispatch scheme 3 MW solar PV.

	Current PV Capacity (3 MW)					
Control Methodology	**Fuel Usage gal p.a.**	**Fuel Usage %**	**O&M $ p.a.**	**O&M %**	**REP %**	**Annual $ Saving**
Existing Dispatch Scheme	2,009,686	100%	$1,887,480	100%	14%	
Proposed Dispatch Scheme	1,991,895	99%	$1,729,860	92%	14%	**$209,469**

Table 8. Performance of existing and proposed low-load dispatch scheme 6 MW solar PV.

	Future PV Capacity (6 MW)						
Control Methodology	**Fuel Usage gal p.a.**	**Fuel Usage %**	**O&M $ p.a.**	**O&M %**	**Excess Energy %**	**REP %**	**Annual $ Saving**
Existing Dispatch Scheme	1,818,893	91%	$1,952,850	100%	3.8%	23%	
Proposed Dispatch Scheme	1,747,453	88%	$1,768,260	91%	2.2%	26%	**$373,616**

4. Discussion

Few approaches to renewable integration acknowledge the resource and capital constraints common within isolated communities, yet both represent significant barriers to uptake of renewable generation. In acknowledging the urgent need for power system decarbonization, low-load diesel application has a role to play in providing greater flexibility to isolated power systems, resulting in improved renewable hosting and acceptance. In this regard, low-load diesel is identified as a low complexity technology solution, offering both hosting capacity relief and battery storage rationalisation. Future research effort should be directed to improved flexibility of dual fuel generation technologies, which remain less compatible to renewable integration than diesel due to increased ignition delay.

5. Conclusions

The presented King Island and Moloka'i case studies represent isolated power systems at differing states of technology progression and refinement. King Island is one of the world's first high-penetration isolated power systems, able to run 100% renewable for extended periods. Unfortunately, the King Island experience has limited commercial relevance given the high cost and complexity of the approach. In this regard low-load diesel has shown to reduce the systems battery requirements by 50%, while reducing annual fuel consumption by 6.3%. The result highlights the gains possible under improved generator flexibility, offering reduced barriers to renewable integration.

For Moloka'i the challenges in progressing past low annual solar PV penetrations are very different. In this environment, low-load diesel is shown to provide near term relief from the hosting capacity constraint currently preventing connection of additional solar PV generation. On Moloka'i, the adoption of low-load applications allows for the interconnection of another 800 kW of approved, but stalled solar PV connections. For a system dependent on residential deployment of renewable generation, the hosting relief offers both commercial and social benefit, reducing community frustration regarding interconnection delay. In addition to the 27% improvement in hosting capacity low-load applications also provided for a 2.7% reduction in operational expenditure. In improving the flexibility of diesel generation on Moloka'i low-load applications provide for improved system efficiency, a recommencement of solar PV connection and a rationalisation of any future battery storage requirement (low-load diesel provided for a 43% reduction in optimal BESS sizing for a hypothetical doubling of renewable capacity). The results identify low-load diesel as a valuable near-term enabling technology, able to deliver significant value with or without BESS integration. Of the available, commercial technologies low-load diesel is unique for its ability to benefit both low and high-penetration isolated power systems. Its accessibility makes it a natural precursor to storage. The results challenge the conventional practice of prioritising high efficiency, low-speed diesel generation as the base load within diesel-based power systems, instead advocating for flexible thermal applications as systems transition towards renewable economies.

Author Contributions: Conceptualization, J.H. and E.S.; methodology, J.H., M.N. and X.W.; software, J.H.; validation, E.S.; formal analysis, M.N. and X.W.; investigation, J.H.; writing—original draft preparation, J.H.; writing—review and editing, X.W. and M.N.; visualization, J.H.; supervision, M.N.; project administration, X.W.; funding acquisition, J.H. All authors have read and agreed to the published version of the manuscript.

Funding: This research was funded by The Australian Research Council, award LP170100879, and the Office of Naval Research Global, award N00014-19-1-2161.

Acknowledgments: The authors would like to acknowledge Richard Rocheleau, from the Hawai'i Natural Energy Institute and Ray Massie from Hydro Tasmania for their assistance during the case study preparation.

Conflicts of Interest: The authors declare no conflict of interest.

References

1. Alves, M.; Segurado, R.; Costa, M. On the road to 100% renewable energy systems in isolated islands. *Energy* **2020**, *198*, 117321. [CrossRef]
2. Aris, A.M.; Shabani, B. Sustainable power supply solutions for off-grid base stations. *Energies* **2015**, *8*, 10904–10941. [CrossRef]
3. Hamilton, J.; Negnevitsky, M.; Wang, X. Economics of renewable energy integration and energy storage via low load diesel application. *Energies* **2018**, *18*, 1080. [CrossRef]
4. Hong, Y.Y.; Lai, Y.M.; Chang, Y.R.; Lee, Y.D.; Liu, P.W. Optimizing capacities of distributed generation and energy storage in a small autonomous power system considering uncertainty in renewables. *Energies* **2015**, *8*, 2473–2492. [CrossRef]
5. Hamilton, J.; Negnevitsky, M.; Wang, X. Economic rationalization of energy storage under low load diesel application. *Energy Procedia* **2017**, *110*, 65–70. [CrossRef]
6. Tsai, C.T.; Beza, T.M.; Wu, W.B.; Kuo, C.C. Optimal configuration with capacity analysis of a hybrid renewable energy and storage system for an island application. *Energies* **2020**, *13*, 8. [CrossRef]
7. Hamilton, J.; Negnevitsky, M.; Wang, X.; Lyden, S. High penetration renewable generation within Australian isolated and remote power systems. *Energy* **2019**, *168*, 684–692. [CrossRef]
8. Corsini, A.; Tortora, E. Sea-water desalination for load levelling of gen-sets in small off-grid islands. *Energies* **2018**, *11*, 2068. [CrossRef]
9. Tafech, A.; Milani, D.; Abbas, A. Water storage instead of energy storage for desalination powered by renewable energy–King Island case study. *Energies* **2016**, *9*, 839. [CrossRef]
10. Hansen, K.; Breyer, C.; Lund, H. Status and perspectives on 100% renewable energy systems. *Energy* **2019**, *175*, 471–480. [CrossRef]
11. Semshchikov, E.; Negnevitsky, M.; Hamilton, J.M.; Wang, X. Cost-efficient strategy for high renewable energy penetration in isolated power systems. *IEEE Trans. Power Syst.* **2020**, *1*. [CrossRef]
12. Gupta, C. Sustainability, self-reliance and aloha aina: The case of Molokai, Hawai'i. *Int. J. Sustain. Dev. World Ecol.* **2014**, *21*, 389–397. [CrossRef]
13. Tasmanian Economic Regulator. Energy in Tasmania Report 2018–2019. 2020. Available online: https://www.economicregulator.tas.gov.au/Documents/Energy%20in%20Tasmania%20Report%202018-19%2020%20210.pdf (accessed on 14 April 2020).
14. Clean Energy Council. Clean Energy Australia Report. 2020. Available online: https://assets.cleanenergycouncil.org.au/documents/resources/reports/clean-energy-australia/clean-energy-australia-report-2020.pdf (accessed on 11 April 2020).
15. Alternative Technology Association. Windpower on Rottnest Island. *Soft Technol. Altern. Technol. Aust.* **1983**, *13*, 11–13.
16. Hawaii State Energy Office. *Hawaii Energy Facts and Figures*; Hawaii State Energy Office: Honolulu, HI, USA, 2019; pp. 33–36. Available online: https://energy.hawaii.gov/wp-content/uploads/2019/07/2019-FF_Final.pdf (accessed on 13 April 2020).
17. Gamble, S. Approaches and lessons from King Island and Flinders Island hybrid projects. In Proceedings of the Informa Remote Area Power Conference, Sydney, Australia, 15 June 2016.
18. Fairley, P. Customers seek 100-percent-renewable grids. *IEEE Spectr.* **2017**, *54*, 12–13. [CrossRef]
19. Reynolds, C. 100% renewable energy Moloka'i. In Proceedings of the Isolated Power System Connect, Maui, HI, USA, 17 October 2018. Available online: http://www.ipsconnect.org/ipsconnect2018/program (accessed on 23 April 2020).
20. Rocheleau, R. Hawaii clean energy transformation policy, challenges, and opportunities. In Proceedings of the Isolated Power System Connect, Maui, HI, USA, 17 October 2018; Available online: http://www.ipsconnect.org/ipsconnect2018/program (accessed on 23 April 2020).
21. Stein, K.T.; Musser, K.; Rocheleau, R. Evaluation of a 1 MW, 250 kW-hr battery energy storage system for grid services for the Island of Hawaii. *Energies* **2018**, *11*, 3367. [CrossRef]
22. Manwell, J.F.; McGowan, J.G. A combined probabilistic/time series model for wind diesel systems simulation. *Sol. Energy* **1994**, *53*, 481–490. [CrossRef]
23. Graham, V.; Hollands, K. A method to generate synthetic hourly solar radiation globally. *Sol. Energy* **1990**, *44*, 333–341. [CrossRef]

24. Manwell, J.F.; McGowan, J.G. Lead acid battery storage model for hybrid energy systems. *Sol. Energy* **1993**, *50*, 399–405. [CrossRef]
25. Lambert, T.; Gilman, P.; Lilienthal, P. Micropower system modeling with HOMER. In *Integration of Alternative Sources of Energy*; John Wiley and Sons: Hoboken, NJ, USA, 2006.
26. Hamilton, J.; Negnevitsky, M.; Wang, X. The potential of low load diesel application in increasing renewable energy source penetration. *Cigre Sci. Eng.* **2017**, *8*, 49–59.
27. Welz, R. Low Load Operation for s1600 gendrive engines. OE Development Newsletter Power Generation, MTU Friedrichshafen, Application Newsletter 15-005, 21st April, Freidrichshafen. 2015.
28. Brooks, P. Limitations on low load operation for fixed speed engines. In *Cummins Marine Application Bulletin*; MSB No. 2.05.00-02172005; Cummins: Columbus, OH, USA, 2014.
29. Hamilton, J.; Negnevitsky, M.; Wang, X. Low load diesel perceptions and practices within remote area power systems. In Proceedings of the International Symposium on Smart Electric Distribution Systems and Technologies (EDST), Vienna, Austria, 8–11 September 2015; pp. 121–126.
30. Hamilton, J.; Negnevitsky, M.; Wang, X.; Tavakoli, A. No load diesel application to maximise renewable energy penetration in off grid hybrid systems. In Proceedings of the Cigre Session, Paris, France, 21–26 August 2016; pp. 1–10.
31. Hawaiian Electric. *100% Renewable Energy Infrastructure Plan Moloka'i*; Hawaiian Electric: Honolulu, HI, USA, 2017.

Article

Investigating the Interrelationships among Occupant Attitude, Knowledge and Behaviour in LEED-Certified Buildings Using Structural Equation Modelling

Mehreen Saleem Gul [1,*] and Elmira NezamiFar [2]

1 Institute of Sustainable Building Design, Heriot-Watt University, Edinburgh EH14 4AS, UK
2 Centre for Construction and Engineering Technologies, George Brown College, Toronto, ON M5R 1M3, Canada; elmira.nezamifar@georgebrown.ca
* Correspondence: M.gul@hw.ac.uk; Tel.: +44-131-451-4082

Received: 13 May 2020; Accepted: 12 June 2020; Published: 18 June 2020

Abstract: The proliferation of residential building energy consumption and CO_2 emissions has led many countries to develop buildings under the green rating systems umbrella. Many such buildings, however, fail to meet their designed energy performance, which is possibly attributable to occupant behaviour and unforeseen building usages. The research problem lies in the fact that occupant environmental behaviour is a complex socio-cultural-technical issue that needs to be addressed to achieve the desired energy savings. This study is novel as it investigates complex interrelationships between many observed and unobserved variables using data from four LEED-certified multi-residential buildings in the United Arab Emirates. Structural Equation Modelling was used to analyse the impact of three unobserved/latent variables: occupant environmental Attitude, Knowledge and Behaviour (AKB) with respect to occupant energy consumption, based on measured/observed variables. Although our Goodness-of-Fit values indicated that we achieved a good model fit, the interrelationship between Knowledge and Behaviour (p = 0.557) and between Attitude and Behaviour (p = 0.931) was insignificant, as the p-values > 0.05. The key study outcomes were: (i) providing information alone could not motivate people towards environmentally friendly behaviour; (ii) even changes in their attitude, belief and lifestyle were not significantly related to their behaviour, as the interrelationships among occupant environmental AKB were not significant; and (iii) knowledge and attitude change should be combined with other motivational factors to trigger environmentally friendly actions and influence behaviour.

Keywords: energy consumption; Structural Equation Modelling (SEM), occupant environmental attitude, Knowledge and Behaviour

---

1. Introduction

Buildings constitute a very high percentage of energy consumption compared to other sectors of the economy. As cited by Ji and Chan (2019), on average, the total energy consumption globally from the residential buildings sector is around 20% which may reach 30% by 2040 due to increasing population, economic activities and the improved standards of living [1]. This situation has led many countries to formulate energy policies that help to reduce energy consumption and ultimately CO_2 emissions. Many countries have begun to adopt mandatory green requirements for their building developments, and green rating systems have become increasingly widely adopted worldwide. Some of the most widely used rating systems include: (i) Leadership in Energy and Environmental Design (LEED), (ii) Building Research Establishment Environmental Assessment Methodology (BREEAM), and (iii) Green Globe Canada [2]; while other well-recognised rating systems include: (iv) Green Star Australia,

(v) Building Environmental Performance Assessment Criteria (BEPAC) Canada, (vi) GB tool Korea, and (vii) Comprehensive Assessment System for Built Environment Efficiency (CASBEE) in Japan [3]. The green rating systems mentioned above have a very limited focus on occupant behaviour monitoring systems during a building's operational phase. Evidence exists [4,5] in support of the fact that by improving occupant behaviour, energy consumption can be reduced by 8–15% in all types of buildings, resulting in lower carbon emissions. The importance of environmentally friendly behavioural improvement is evident from the fact that many buildings using new technology-oriented systems fail to meet their 'as designed' performance expectations [6]. Large discrepancies exist between as-designed (predicted) and real building energy performances (performance gap), typically averaging around 30% [7]. At least some of this performance gap is attributable to unforeseen usages by occupants of these green/LEED-certified buildings and their equipment.

It has been known for a long time that occupant behaviour can greatly influence energy use in buildings [8]. As stated by Nguyen and Aiello [9], the ways in which occupants interact with a building have shown to exert large impacts on heating, ventilation and air-conditioning (HVAC) demand and building controls. According to Schipper et al. (1989) [10] where 50% of energy use in homes is attributed to the intrinsic building shell, HVAC, lighting, and electronics, the rest occurs due to the occupant interaction with these systems. In their study in mid-1990s, FSEC determined the magnitude of occupancy-related effects by examining energy usage in ten identical homes in Florida [11]. With the same number of occupants and identical appliances and equipment, energy use varied by 2.6 to 1 from the highest to the lowest consumer with a standard deviation of around 13 kWh/day—32% of the mean. On further examination FSEC found that while the electrical consumption of appliances like refrigerators were remarkably similar, air conditioning e.g., varied by 5:1 from highest to lowest. The measurements of interior temperature displayed huge differences due to differing thermostat behaviour. In another study [11,12], the utility bills of eleven similarly efficient solar homes in Sacremento were compared to other non-solar homes in the same community, to show large variations in annual energy use in the solar homes. In comparison with the most frugal home, where solar electricity generation was more than the consumption, the highest-consuming solar home used almost twice as much electricity as the average energy use of non-solar homes. These studies suggest that motivating changes to occupant behaviour could be a powerful measure for achieving energy reductions, especially in more efficient homes with renewable energy features.

Occupant environmental behaviour is therefore one of the major reasons behind the significant uncertainty regarding building energy consumption and the performance gap [13,14]. Dynamic building energy models are now commonly used in academia and industry for a detailed analysis of heating and cooling energy consumption. Although these models interpret thermal behaviour of a building with high precision in relation to its ambient indoor and outdoor environment, the interpretation of occupant behaviour is subjected to relatively simplified input data. Extensive research has been undertaken to evaluate the sensitivity of models to the buildings technical design parameters whereas evaluation of factors such as energy management and energy users behaviour, which play a significant role in influencing building energy consumption, have rarely been performed [15,16]. Several recent studies have focused on measures for achieving highly efficient and comfortable buildings, using the Operative Air Temperature or the PMV and PPD indices as parameters, but sometimes the numerical predictions are different from the real performance of the building, if the behaviour of the occupant is not taken into account [17]. Other studies have revealed that advances in technology and investments alone cannot warrant a low energy future in a building. Zhang et al. (2020) e.g., undertook a survey in Beijing, China highlighting that no significant correlation exists between occupant purchase behaviour of energy efficient equipment against their usage behaviour and confirmed that there was no coherent pattern that could be explained by any single socio-demographic factor [18]. Ashouri et al. (2019) introduced a new ranking procedure for performing comparisons between the occupants of several buildings in order to perform an evaluation of the energy performance of each building in comparison with the others and provide suggestions to occupants for energy conservation in order to improve their rank [19].

The widening gap between real energy consumption and the estimated energy consumption at the design stage demands that behavioural aspects should be taken into account more rigorously than ever [20]. A better understanding of occupant environmental behaviour is therefore needed to improve this uncertainty [21] and to reduce energy consumption by up to 8–15% in all types of buildings [4,5].

The industry contains green building initiatives, such as the LEED and many other good practices around the globe, which are meant to promote sustainability in the built environment by incorporating measures that raise awareness among the public about environmental issues and energy conservation. Most studies on energy consumption of LEED-certified buildings have concluded that an energy performance gap exists between predicted and actual energy consumption [6]. One of the key common factors is that the buildings may not be operated properly in cases where a knowledge gap exists with respect to energy between the industry professionals, the building operators, and the occupants. In addition to these considerations, occupants perform various actions to satisfy their needs in buildings; actions which can negatively affect building energy consumption, because those occupants do not always behave in an environmentally friendly manner to achieve the energy-saving potential of their buildings [22]. This micro-focus has therefore created a research opportunity to investigate LEED-certified residential buildings in use in detail and to explore how we can better understand occupant behaviour through more intensive post occupancy evaluations.

In this paper, we introduce a novel way of using Structural Equation Modelling (SEM) to investigate in detail the interrelationships between latent (unobserved) variables, occupant environmental Attitudes, Knowledge and Behaviour (AKB), based on observed data to understand their true environmental behaviour. Questionnaire data from the residents of four LEED-certified multi-residential buildings in Dubai, United Arab Emirates (UAE) were used for this purpose. Data analysis was conducted through SPSS for the survey questionnaire, with the data later being transferred to Analysis of Moment Structure (AMOS) software in order to develop a measurement and a structural model using the SEM.

The paper is structured as follows: Section 2 provides background and context on the challenges associated with energy conservation for the chosen location, Section 3 describes the research methodology, the SEM technique used and the development and enhancement of the measurement and structural models, Section 4 describes the questionnaire and experimental results, Section 5 discusses the results and Section 6 draws the conclusions.

2. Background and Context

The hot and arid climate of the UAE poses great challenges with respect to reducing energy consumption in buildings. The extremely high insolation and humidity levels together with the lack of consideration given towards energy conservation and green practices over the several years of its early development, identify the UAE as one of the top ten countries in electricity usage and second in carbon dioxide emissions per capita [23]. Over the past 20 years, the UAE has experienced rapid growth, which has resulted in a large and growing stock of modern high-density buildings. Today the UAE has become one of the world's biggest per capita air polluters, and it has been listed as the country with the highest per capita fossil fuel consumption and carbon dioxide emission rates worldwide [24]. In addition, because of increasing tourism, together with average population growth, the UAE's demands on natural resources have also increased in terms of water and energy consumption, in addition to a massive production of waste.

In the UAE, cooling accounts for almost 80% of a building's electricity demand. The outdoor air temperature in the UAE is above 25 °C for 75% of typical working hours, with relative humidity being above 60% for more than 20%, and insolation being in excess of 893 W/m^2 for more than 15% of the year. These environmental conditions necessitate the use of mechanical cooling by air conditioning to maintain internal thermal comfort for the majority of the year [25]. It is reasonable to conclude that the construction industry practices in the UAE were not sustainable when they were created, especially when compared to today. The focus of investors was mainly on obtaining the quickest returns on their investments; a focus that ultimately led to the downfall of the UAE's construction sector. Studies [24,26] have demonstrated the high energy consumption and CO_2 emissions of most existing buildings in Dubai and Abu Dhabi when compared to international benchmarks. Statistics show that 43% of the CO_2 production is due to electricity usage within buildings in the UAE, and only 4% is due to the direct emissions of buildings [26]. The UAE's government has recognised the importance of energy efficiency and has focused on the building sector as the main energy consumer.

In 1991, the UAE established an NGO called the Emirates Environmental Group for the purpose of promoting sustainability in the UAE [27]. Since then, buildings in the commercial stock, in both the growing and newer buildings, have increasingly adopted energy-efficient strategies to address demands for cooling. These changes in newer buildings are influenced by the national drive towards sustainability, and particularly the introduction of green building regulations in 2003, when the Dubai municipality enforced Degree 66 as an energy saving approach. These savings were to be achieved by improving the insulation and glazing systems in building. Subsequently, the Emirates Green Building Council (GBC) was created in 2006 to ensure environmental sustainability in the UAE, and the Estidama programme was established by the Abu Dhabi Planning Council in 2008, involving guidelines for both the design and operation of sustainable buildings [28]. The UAE substantially promoted sustainable development after the 2008 economic crisis in order to bring the construction sector in the UAE into line with international sustainability standards [29] and chose sustainability as an important factor in its bid for EXPO 2020 [30,31]. After winning the bid, the EXPO 2020 sustainability policy states the intention to host one of the most sustainable World Expos in history [32].

Masdar City is one of the most remarkable projects in the UAE, such as; a carbon-neutral and sustainable city powered by renewable energy technologies under the supervision of the government-owned Mubadala Development Company for Abu Dhabi vision 2030. Innovative designs and technologies, such as (i) solar panels, (ii) wind turbines, (iii) recycled glass, (iv) high-temperature plasma torch systems, and (v) non-toxic plastic products, are used to promote a safer environment in the Masdar City Project [33]. Thermal insulation and green building codes have been applied in both Dubai and Abu Dhabi; however, there is no model for analysing the impact of these codes on the reduction of CO_2 emissions [26]. The most popular green agencies for buildings in the United Arab Emirates (UAE) are the LEED, which is mostly used in Dubai, and then BREEAM, which is mostly applied in Abu Dhabi [28].

The existence of green building initiatives such as LEED and many other good practices regulated by the UAE government [28,30,31,34] does not necessarily mean that a building's occupants behave in an environmentally friendly fashion. LEED's rating categories generally encourage sustainable design, health, and economic benefits; however, they do not consider the significance of the human dimensions: capabilities, attitudes, knowledge and behaviour [35,36]. There is clearly a need for further research to clarify this issue in order to bridge the gap between estimated and true energy savings. Jones and Vyas (2008) [37] stated that measuring or verifying the post-occupancy performance of homes will help by increasing the data available for improving on the real performance attributes of green residential buildings. The researchers noted that changes in occupant behaviour could be achieved by addressing everyone's energy consumption awareness, as well as facilitating the occupant group's knowledge and perceptions through advertising, marketing, and other information strategies [38,39]. Therefore, this research explored whether the occupants possessed the knowledge to change and/or improve their environmental behaviour in order to achieve energy savings, and whether their beliefs and attitudes could lead them towards greener behaviour. To fulfil the stated purpose, the interrelationships among occupant environmental Attitudes, Knowledge and Behaviour (AKB) were analysed as described in the later sections.

3. Research Methodology

The research design is summarized in Figure 1 as follows:

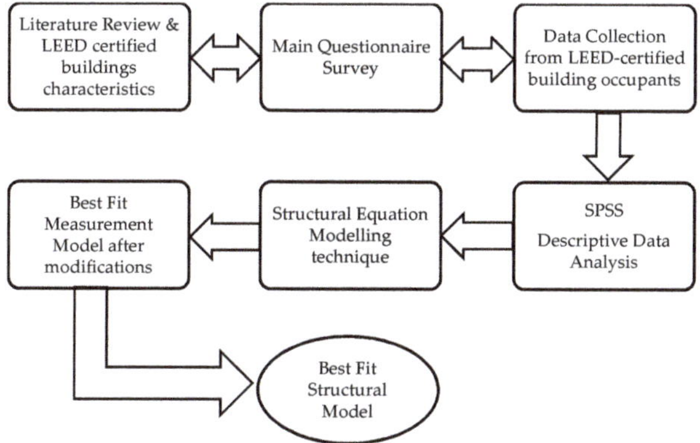

Figure 1. Research design diagram.

3.1. Questionnaire Survey

In order to understand better different factors affecting occupant behaviour and their preferences to behave in a certain way, a questionnaire survey was designed to collect quantitative data needed to confirm the relationship between the observed and latent variables. The design of the questionnaire was based on studies on post occupancy evaluation (POE) survey questions and questions for other similar research studies [40]. Reviewing such resources helped to address the identified questions in a way that was relevant to this research study. There was a total of 31 short questions with multiple choice answers. For simplicity, the structure of the questionnaire was made as easy as possible for the participants to respond to. The questionnaire consisted of five sections, starting with demographic questions followed by questions relating to occupant comfort/satisfaction level and the effectiveness of the management system in relation to the training and knowledge sharing in their buildings. The questions were aimed at gauging the attitudes and knowledge that influence the occupant's environmental behaviour.

The questionnaire was rigorously tested for ease to manage and understandability for the volunteer respondents. Only a few modifications were made after completing the testing procedure. For example, there was no 'Do not know' option for the multiple choice questions, and after observing the lack of answers to some of the questions (knowing that occupants did not know the answer), it was decided to add the option to avoid missing values while analyzing the data in SPSS and AMOS.

3.2. Survey Sampling

By the time of setting the research study, there were 15 LEED-certified residential multi-family buildings and 14 villas with approximately 1724 units with less than a 60% occupancy rate. By speaking with building operators and the USGBC, it was found that there were approximately 1034 occupants in those LEED-certified units in the UAE in 2014 [41]. The required sample size was then calculated with a 5% margin of error, 95% confidence level, and 50% sample proportion as follows:

$$\begin{aligned}&\text{Sample Size (X)} = \text{Distribution}/(\text{Margin of Error}/\text{Confidence Level Score})^2\\&\text{Sample Size (X)} = 0.5/(0.05/0.95)^2\\&\text{Sample Size (X)} = 180.500202\end{aligned} \quad (1)$$

By putting Sample Size (X) in the True Sample formula:

$$n = X \times N/(X + N - 1), \quad (2)$$

where 'n' is True Sample, X is Sample Size and N is Population,

$$\text{True Sample (n)} = 180.500202 \times 1034/(180.500202 + 1034 - 1) = 154$$

The required response rate thus was 154 to reach the minimum acceptable threshold/True Sample. The questionnaire was distributed via email, as well as hand delivered. Respondents were reminded every three weeks through follow-up emails, or notes through their door, in order to improve the response rate. If a potential candidate was reminded two times, but still declined to respond then that occupant was removed from the list of potential participants.

3.3. Building Selection

Most occupied LEED-certified buildings in 2014 were in 'Dubai International City', from which four LEED-certified multi-residential buildings were recruited (Figures 2–5). The key specifications of these buildings are provided in Table 1. PR I and PR II were completed in March 2011 and LEED certified in June and October 2011, respectively. TC was completed in July 2011 and certified in August 2012, and HDS SS II was completed in September 2012 and certified in September 2013. Out of the 628 units of the four LEED-certified buildings, there were 265 occupied units at the time of the survey. Therefore, the questionnaire was distributed to a total of 265 occupants residing in those units (flats/apartments). If more than 154 occupants out of 265 had not participated, then the researchers would have had to target more residential units to reach the minimum acceptable threshold (True Sample). A total of 203 occupants responded to the survey with valid answers, resulting in a response rate of 76.6%. Although the response rate is good, the authors acknowledge that their sample is limited to Dubai residents only, which might have incurred bias in terms of environmental concerns, age, etc. Therefore, future research is recommended, using datasets from a wide range of demographics and geographical regions.

Figure 2. Prime Residency I (PR I), International City, Dubai, UAE.

Figure 3. Prime Residency II (PR II), International City, Dubai, UAE.

Figure 4. Trafalgar Central (TC), International City, Dubai, UAE.

Figure 5. HDS Sun Star II (HDS SS II), International City, Dubai, UAE.

Table 1. Information regarding 4 LEED-certified residential buildings.

Fast Facts	Prime Residency I (PR I) 164 Units	Prime Residency II (PR II) 164 Units	Trafalgar Central (TC) 160 Units	HDS Sun Star II (HDS SS II) 140 Units
LEED Certification	SILVER New Construction (NC) V2.2	GOLD New Construction (NC) V2.2	SILVER New Construction (NC) V2.2	SILVER New Construction (NC) V2.2
Area (Square Feet)	251,176 sq. ft.	251,176 sq. ft.	214,059 sq. ft.	176,485 Sq. ft.
Estimated Savings & Benefits	17.5% less energy use 35.4% less potable water use 20.9% materials use with recycled content 43.9% local materials use	38.6% less energy use 35.4% less potable water use 20.9% materials use with recycled content 43.9% local materials use	17.9% less energy use 41.3% less potable water use 34.38% materials use with recycled content 41.24% local materials use	21.3% less energy use 31.8% less potable water use 21.0% materials use with recycled content 13.2% local materials use

3.4. Data Analysis

The collected data was first analysed using the SPSS statistics software (version 22) to obtain descriptive statistics, frequencies, and means, after which the data was transferred to AMOS for deeper analysis using Structural Equation Modeling (SEM) techniques. Descriptive statistics were used as a set of descriptive coefficients to summarise a given data set, which was a representation of the entire population. The mean rating statistical technique was selected to analyse participants' ratings of the importance of different factors when choosing their homes, by using the numerical values assigned to each factor to compute their mean scores.

The SEM approach was chosen as it was the most appropriate data analysis method for this part of the study. Larger sample sizes (100–400) generally regarded as acceptable for SEM analysis among researchers [42]. Therefore, the sample size of 203 (survey participants) in the current study was considered to meet the threshold of acceptability. Among all the available software, AMOS was chosen, as it is the most recent statistical package which has a user-friendly graphical interface, and it has become popular as a simpler way of specifying structural models.

The SEM is a forecasting method which can be used in a variety of contexts and offers a confirmatory, rather than an exploratory approach to the data analysis. Most multivariate procedures (e.g., exploratory factor analysis) are essentially descriptive in nature whereas SEM allows analysis of data for conclusive purposes [42]. SEM has recently become an essential and influential statistical method in social science research [43]. Ji and Chan [1] described SEM as a second-generation multivariate analysis technique which combines the functions of exploratory factor analysis and linear regression analysis to achieve the assessment of both the measurement model and structural model simultaneously. SEM offers flexibilities by encompassing various formats and large numbers of variables with fewer limitations e.g., to the sample size and data normality [44–46]. SEM has gained popularity in many research areas, e.g., strategic management [47], information systems [48], business management [49,50], tourism management [51], accounting [52], technology adoption by the construction industry [53], and marketing [54].

A limitation of traditional multivariate procedures is their incapability of either assessing or correcting for measurement error, SEM on the other hand presents explicit estimates of these error variance parameters. It offers a powerful substitute to multiple regression, path analysis, factor analysis, time series analysis and analysis of covariance. SEM became popular for non-experimental research, where methods for testing theories were not well developed or where ethical considerations make experimental design unfeasible [55,56]. According to Hair et al. (1998) [57], SEM should usually be developed through several stages; first, to define structural components to identify the measurement components which deal with the relationships among the unobserved/latent variables and their indicators/observed variables, then to set up a model specification (hypothetical model) based on the aim of the research, and subsequently to evaluate the model estimates in order to validate the structural model variables and finally to modify the model based on potential changes. By using the confirmatory factor analysis (CFA) approach, SEM makes it possible to review the interrelationship between observed variables and their underlying latent variables. This technique was then used to test the interrelationships among the latent variables affecting occupant behaviour. Where goodness-of-fit is satisfactory, the model shows that there are interrelationships among variables, but where this is inadequate, then the interrelationships among the variables are rejected [58]. At least three observed variables/indicators are recommended and a common practice whereas, problem exists with two or one observed variable as the measurement error cannot be modelled [59]. If models use only two observed variables per latent variable, they are more likely to fail, and therefore error estimates might be unreliable.

For the present study, SEM was therefore used to develop a model to quantify complex relations between the environmental Attitude, Knowledge and Behaviour (AKB) of the occupants, as shown in Figure 6. Such a model refers to implicit or explicit models that relate the latent variables to their observed variables. The measurement model, shows the relations between the latent variables AKB and their observed variables, where the structural model presents the interrelationships among the latent variables (AKB) only. Observed variables within questionnaires included five different sections: (i) building occupant backgrounds, (ii) knowledge, (iii) attitude, (iv) behaviour, and (v) satisfaction level. Three sections of the questionnaire which defined ATTITUDE, KNOWLEDGE and BEHAVIOUR known as AKB in this research study, were chosen for further analysis. The answers to each question were considered to be observed variables while the whole AKB cluster were labelled latent variables. It is the significance of the interrelationships between them that should be measured, analysed and modelled.

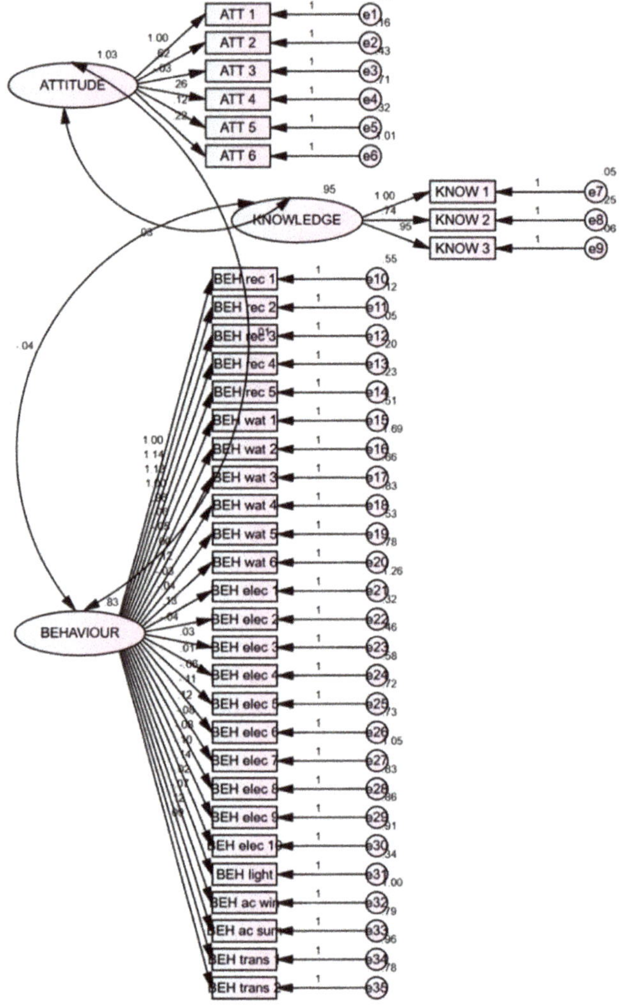

Figure 6. Conceptual measurement model of interrelation between building occupants environmental AKB.

4. Results

The results of the questionnaire analysis in SPSS prior to conducting the SEM technique indicated that most occupants were young, the majority averaging between 30 and 49, with college/university degrees and a good knowledge and understanding of sustainability. The majority were renters with an average occupancy number of 2–4 persons. More than 65% of the respondents expressed concerns about the environment and considered climate change to be a global threat. On the other hand, the findings showed that while the majority were concerned about the environment and were aware of sustainability and climate change issues, only relatively few behaved in an environmentally friendly fashion in their daily lifestyle. The comments from the questionnaire indicated that although the majority of the respondents were aware of climate change and they claimed that they considered climate change to be a major or minor threat, and also confirmed the need for energy conservation, they still chose to behave in an un-environmentally friendly fashion; for example, they responded that they run the washing machine at a higher setting rather than using it at the eco mode, or they run the tap while brushing their teeth, etc. It is fair to conclude that their attitudes and values were not leading them to behave in an environmentally friendly manner. This outcome confirms the revelations from the literature review [60], where the authors noted there were intervening constructs between attitude and behaviour. They mentioned knowledge as an intervening factor, and this research study adds to this finding, as the knowledge should inform and motivate occupants to engage in environmentally friendly behaviour.

Occupants explained that the main factor for moving into the building was the location, which provides easy access to their workplaces. Energy efficiency remains a minor consideration for consumers when choosing a home, with the majority of respondents identifying it as the least important factor out of six choices. In general, occupants displayed some positive attitudes towards environmental behaviour, such as energy and water conservation, but not for recycling, and especially not for alternative transportation. Recycling behaviour, on the other hand, was the most considered behaviour, in comparison to other environmentally sensitive conduct. Several occupants constantly displayed positive attitudes towards recycling due to their knowledge, previous habits, culture, and beliefs, and emphasised how effortless it was to recycle. The findings regarding using public transportation show that occupants relied on their vehicles for grocery shopping even during the wintertime and nice weather. Furthermore, none of the occupants mentioned that they owned a hybrid car.

The majority of occupants confirmed that their cooling system was responsive enough for their needs, so they were able to enjoy a satisfactory thermal comfort level. However, the findings showed that AC working hours, especially during summer days, were very high, and that this caused high energy bills. The other highly rated factor in satisfaction was related to privacy and security; in particular, the latter quality applied not only to their buildings, but to the whole of the UAE. Most of the respondents agreed that there was sufficient light in their homes during daylight hours in order not to need artificial light. There were some exceptions related to lack of daylight caused by inappropriate building design. The occupants agreed that maintenance and operation was of an acceptable level, with some occupants complaining about the quality of the building materials used and the finishes. Dissatisfaction was expressed with respect to high energy bills, and purchase and rental cost, and when asked if they had sufficient garden space and recreational areas, the majority of them were strongly dissatisfied.

Although many occupants received instruction manuals and/or training on how to operate the technologies inside their apartments, the results demonstrated that the quality of information was inconsistent, and was often inadequate or overly complicated, rendering the information incomprehensible, and therefore of little value to the occupants. It was also confirmed that no post-occupancy evaluations were conducted, and no information/feedback processes were implemented to raise awareness among occupants about the environmental outcomes of their behaviour.

The findings highlighted that green buildings require a high level of occupant and operator engagement and understanding in order for them to remain holistically green. Low levels of education, motivation and coordination lead to minimal levels of environmentally focused activity from the building occupants. The survey findings suggested that training and education (Knowledge) driven by motivational factors such as i) creating a good socio-cultural environment, and ii) offering financial incentives are key concerns for achieving sustainable development. Matching technology, management sophistication, understanding roles, social organisation, and interactions among building occupants and operators, together with economic incentives, combine to constitute a major avenue through which proper environmental behaviours can be encouraged.

The authors found inconsistency in occupant AKB-related results, and therefore they decided to further investigate the interrelationships between AKB using the SEM technique as described below.

4.1. Conceptual Measurement Model

The six questions under the attitude-related survey questionnaire were chosen as observed variables by the authors for ATTITUDE, twenty-six questions were included for the behaviour-related survey questionnaire as observed variables for BEHAVIOUR, and three questions related to the provided information and guidelines were chosen for KNOWLEDGE.

The conceptual (hypothesized) measurement model generated in AMOS is presented in Figure 6. The model shows the interrelationships among the latent/unobserved variables (AKB) and their indicators/observed variables (questionnaire data) for the purpose of assessing goodness of fit and/or validity. The unobserved/latent variable is linked to an observed variable, in order to make its measurement possible. Assessment of 'ATTITUDE' (ATT) constitutes the direct measurement of six observed variables obtained through the questionnaire survey; similarly, 'BEHAVIOUR' (BEH) is measured through 26 variables (based on recycling, water use, electricity use, lighting use, summer/winter use and transport) and 'KNOWLEDGE' (KNOW) is measured through three observed variables. Further analysis was performed through SEM in AMOS, and was categorised and coded as given below:

ATTITUDE

- ATT 1: View on climate change
- ATT 2: Belief about the impact of energy use on the environment
- ATT 3: Current lifestyle related to the environment
- ATT 4: Environmentally friendly lifestyle changes and the comparison between now and 4 years ago
- ATT 5: Attitude and belief about green buildings
- ATT 6: Considering the term LEED-certified while choosing the home

BEHAVIOUR

- BEH rec 1: Occupant behaviour towards recycling papers
- BEH rec 2: Occupant behaviour towards recycling plastic pieces
- BEH rec 3: Occupant behaviour towards recycling glass
- BEH rec 4: Occupant behaviour towards recycling metal pieces
- BEH rec 5: Occupant behaviour towards recycling carton boxes
- BEH wat 1: Using a washing machine economically
- BEH wat 2: Using a dishwasher economically
- BEH wat 3: Using less water in toilets
- BEH wat 4: Pressing both buttons on WC flush
- BEH wat 5: Taking showers instead of bathing
- BEH wat 6: Turning tap off when brushing teeth
- BEH elec 1: Leaving appliances on standby mode
- BEH elec 2: Turning off lights if they're not needed
- BEH elec 3: Using low energy light bulbs
- BEH elec 4: Using low energy labelled appliances
- BEH elec 5: Setting the thermostat for air conditioning
- BEH elec 6: Keeping AC off when windows are open
- BEH elec 7: Keeping windows open during summer
- BEH elec 8: Keeping windows open during winter
- BEH elec 9: Closing window shades/blinds
- BEH elec 10: Controlling doors/windows airtightness
- BEH light: Hourly usage of artificial lighting in a day

- BEH ac win: Hourly working of AC in a winter day
- BEH ac sum: Hourly working of AC in a summer day
- BEH trans 1: Walking or cycling to your work/supermarket
- BEH trans 2: Using public transportation

KNOWLEDGE

- KNOW 1: Day-to-day energy usage guide
- KNOW 2: Operation and maintenance guide
- KNOW 3: Emergency cases guide

Figure 6 shows that the CFA model focuses solely on the interrelation between AKB factors and their measured variables. The model determines the goodness-of-fit between the factors in the hypothesised model and the sample data. The factor loading between each latent variable and its observed variables is important to be higher, i.e., ATTITUDE and its observed variables ATT3, ATT4, ATT5 and ATT6, with 0.03, 0.26, 0.12 and 0.22, respectively, are very low, which might be problematic, as they should be closer to 1 in order to achieve goodness of fit.

$$t < 12 \times s(s + 1)$$

The above equation, if confirmed, indicates that the model in Figure 6 is over identified:
t = items to be identified= 70 (35 'e' + 32 factor loading + 3 latent variables)s = number of observed variables (35)

$$70 < 630$$

Based on the above result, it is confirmed that the model (Figure 6) is over identified, meaning that we have more than enough observed variables (35) to identify unobserved items (70), and therefore we have the possibility of eliminating some if we need to, in order to achieve the best model fit.

4.2. Conceptual Measurement Model Evaluation

The list given below explains the acceptable and good fit data ranges [58,61] that need to be achieved in order to confirm that a model is a good fit:

- Ratio of minimum discrepancy to degrees of freedom (CMIN/DF)—This adjusts the chi-square by calculating the ratio of the minimum discrepancy to degrees of freedom. It ranges between 1 to 2 where values closer to 1 indicate better fit.
- Goodness of fit index (GFI)—this determines whether the maximum likelihood estimate of the hypothesised model fits to the data set. It ranges between 0 to 1, GFI > 0.9 means satisfactory fit
- Adjusted goodness of fit (AGFI) favours parsimony, AGFI > 0.90 indicates a good fit.
- Incremental fit index (IFI)—This is the ratio of the difference between the hypothesised and baseline model degrees of freedom and discrepancy. It ranges between 0 and 1 where higher values indicate better fit.
- Normed fit index (NFI)—NFI > 0.9 means satisfactory fit, and values greater than 0.80 suggest a good fit and indicates that the model of interest improves the fit by 80% relative to the model.
- Non-normed fit index (NNFI) is preferable for smaller samples. NNFI is also called the Tucker-Lewis index (TLI). TLI > 0.9 indicates a satisfactory fit; TLI compares degrees of freedom and discrepancy between baseline model and those of the hypothesised model. It ranges between 0 and 1 where larger values indicate better fit.
- Comparative fit index (CFI)—This compares the fit of a baseline model to the data with the fit of the hypothesised model to the same data. It ranges between 0 and 1, where larger values indicate a better fit. CFI > 0.9 means satisfactory fit.

- Relative fit index (RFI), also known as RHO1, is not guaranteed to vary from 0 to 1. RFI close to 1 indicates a good fit.
- Root mean square residual (RMR) computes the residual differences between model prediction and data set, and it also takes the square root of the result. It ranges from 0 to 1, with smaller values indicating better fit.
- Root mean square error of approximation (RMSEA)—this is a measure of how close/approximate the fit of population data is with the model. <0.05 good fit and <0.08 acceptable fit. LO 90 and HI 90 include the lower and upper limits of a 90% confidence interval for the population
- The parsimony ratio (PRATIO) is the ratio of the degrees of freedom in the model to degrees of freedom. It is not a goodness-of-fit test in itself, but is employed in goodness-of-fit measurements like PNFI and PCFI.
- The parsimony goodness of fit index (PGFI), is a variant of GFI that penalises GFI by multiplying it by the ratio formed by the degrees of freedom in the model divided by degrees of freedom in the independence model.
- The parsimony normed fit index (PNFI) is equal to the PRATIO times NFI.
- The parsimony comparative fit index (PCFI) is defined as PRATIO multiplied with CFI. PCFI closer to 1 is a better fit.
- PCLOSE tests the null hypothesis that RMSEA is not greater than 0.05. If PCLOSE is less than 0.05 the hypothesis is rejected to conclude that the computed RMSEA is greater than 0.05 which indicates the lack of a close fit.

The significance of the interrelationships among variables in the measurement/hypothesised model was tested in AMOS software in order to review the reliability and commonality of such a model. The results are presented in Table 2.

Table 2. Conceptual measurement model fit.

CMIN	CMIN 3445.947	DF 557	P 0.000	CMIN/DF 6.187	
RMR, GFI	RMR 0.133	GFI 0.478	AGFI 0.410	PGFI 0.423	
Baseline Comparisons	NFI 0.375	RFI 0.332	IFI 0.417	TLI 0.372	CFI 0.412
Parsimony-Adjusted Measures	PRATIO 0.936	PNFI 0.351	PCFI 0.386		
RMSEA	RMSEA 0.160	LO 90 0.155	HI 90 0.165	PCLOSE 0.000	

The list below checks whether the parameters in Table 2 related to the data for Figure 6 are within range and are a good fit, as explained above:

- In Table 2 CMIN/DF is 6.187, which is not a good fit, upper threshold is 5.
- GFI is 0.478, which ranges between 0 and 1, but is still not completely satisfactory.
- AGFI is 0.410 and is not a good fit.
- IFI is 0.417 is within the range, but it is better to be closer to 1 for a better fit.
- NFI is 0.375 and it is not satisfactory.
- TLI is 0.372, which is within the range, but it is not the best fit and is not satisfactory.
- CFI is 0.412, which is within the range, but it is not the best fit.
- RFI is 0.332, and it is not the best fit, although it is within the range.
- RMR is 0.133, and it is within the range.

- RMSEA is bigger than 0.08, at 0.160, which is not even an acceptable fit. In Table 2, LO 90 is 0.155 and HI 90 is 0.165 which constitute the lower and upper limit of a 90% confidence interval for the population value of RMSEA
- PRATIO is 0.936, which would be a better fit with a smaller value.
- PGFI is 0.423, which is not a good fit.
- PNFI is 0.351, where a value closer to 1 would be a better fit.
- PCFI is 0.386, which is not satisfactory.
- PCLOSE is 0, which is less than 0.05, RMSEA is 0.160, which is greater than 0.05; therefore the model is not a close fit (Figure 6).

All of the figures in Table 2 indicate that Figure 6 is not a well-fitting measurement model, meaning that none of the observed variables/indicators have strong interrelationships with their own latent variables AKB.

4.3. Conceptual Measurement Model Modification

Based on the results given above, the measurement model was not a good fit, and needed to be modified. To achieve the best fitting measurement model, three main steps [62] were carried out, as follows:

- The first step was to delete paths which had very low factor loadings,
- The second step was to covary variables based on the modification indices (MI) in Table 3,
- The third step was to eliminate the observed variables with very high values in the standardised residual correlation matrix,

Table 3. Modification Indices ≥11 for Step 2 modifications of observed variables interrelationships

e13 ↔ e14 = 75.421	e11 ↔ e14 = 14.860
e10 ↔ e13 = 12.434	e10 ↔ e11 = 34.255
e24 ↔ e27 = 34.311	e23 ↔ e27 = 11.685
e23 ↔ e24 = 84.190	e21 ↔ e27 = 119.003
e21 ↔ e24 = 15.784	e14 ↔ e21 = 11.573

The path coefficient and GOF sometimes reveal the need to modify models in SEM, which can result in the selection of the best fitting model falling within theoretical expectation and satisfying the GOF measures [42].

STEP 1:

In the first step, some of the observed variables with a very low factor loading were eliminated from the measurement model as follows:

- From ATTITUDE: ATT 3, ATT 5, ATT 6 were eliminated; ATT 4 (0.26) was kept, as it was an important observed variable for supporting the latent variable.
- From BEHAVIOUR: BEH wat 1, BEH wat 2, BEH wat 5, BEH wat 6, BEH elec 2, BEH elec 5, BEH elec 6, BEH elec 8, BEH elec 9, BEH elec 10, BEH light, BEH ac win, BEH ac sum, BEH trans 2 were eliminated.

This was due to the very low factor loading shown by the arrow between the latent variable and the observed variables in Figure 6. Figure 7 shows the revised measurement model, including the observed variables with higher factor loadings after Step 1, although this might not be the best measurement, and further steps as described above should be carried out.

$$t < 12 \times s(s + 1)$$

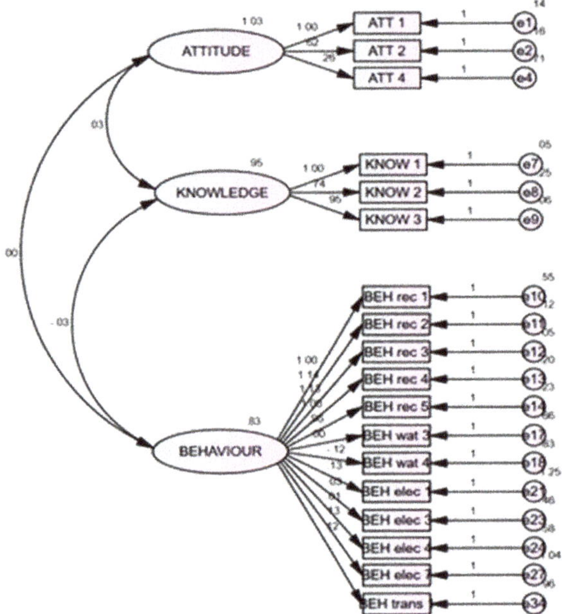

Figure 7. Conceptual measurement model of interrelation between building occupant environmental AKB after Step 1.

The above equation, if confirmed, indicates that the model in Figure 7 is over identified:
t = items to be identified= 36 (18 'e' + 15 factor loading + 3 latent variables)
s = number of observed variables (18)

$$36 < 171$$

Based on the above result, it is confirmed that the model (Figure 7) is over identified, meaning that we have more than enough observed variables (18) to identify the unobserved items (36).

Modification Indices (MI) are often used to modify models in order to achieve a better fit, but this process should be carried out carefully and with theoretical justification [58]. For the MI generated in Step 1, the threshold was set at 10, and the MI valued as equal to/greater than 11 were selected for Step 2 as shown in Table 3.

STEP 2:

In this step the model fitting results for Figure 7 were reviewed in light of the selected modification indices (Table 3) from Step 1. The authors considered only those observed variables to be covaried that corresponded to the MI that were equal to/greater than 11.

Covarying can be seen in Figure 8 as small curved two-way arrows between 'e' (error variance) of observed variables in the same factor that were equal to or more than 11. All of the errors (e) were from 'Behaviour', and they represented covariates or correlations between pairs of variables. For example, the MI between 'e13' and 'e14' in Table 3 is 75.421; therefore, it is above the threshold, which was set at 10, and they should be covaried, as shown in Figure 8.

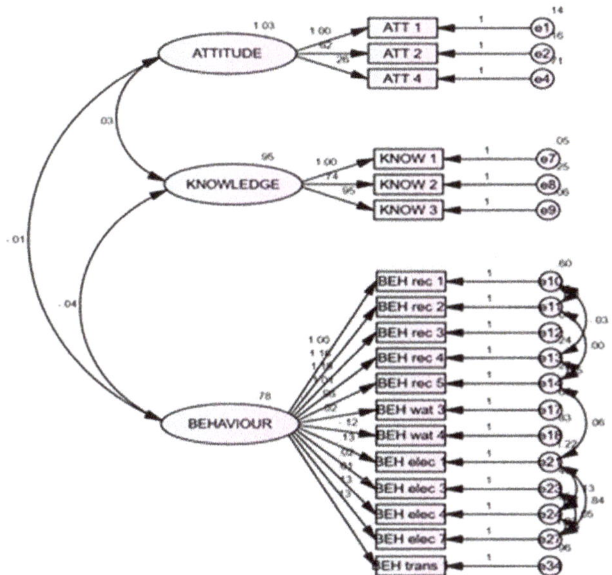

Figure 8. Conceptual measurement model of interrelations among building occupants environmental AKB after Step 2.

Figure 8 is the modified measurement model after Step 2, covarying for all the 'e' that are shown in Table 3, which is not yet the best model fit, due to the presence of the low factor loadings shown between each latent variable and their observed variables. There is still some factor loading below 0.1 shown by the arrows in Figure 8. For example, the arrow going from the latent variable 'Behaviour' to its observed variable 'BEH wat 4' was very low at −0.12, and the arrow to 'BEH wat 3' was 0.02.

STEP 3:

The third step was to go through the standardised residual covariance in Figure 8 after correlating errors and deleting some of the observed variables above 0.1, shown on the arrows from Behaviour to its indicators/observed variables. Different trials were conducted, and the best outcome was to eliminate the following observed variables: BEH wat 3, BEH wat 4, BEH elec 3, BEH elec 4, BEH elec 7, and BEH trans 1 in BEHAVIOUR. The resultant best fitting measurement model can be seen in Figure 9.

$$t < 12 \times s(s + 1)$$

The above equation, if confirmed indicates that the model in Figure 9 is over identified:
t = items to be identified= 24 (12 'e' + 9 factor loading + 3 latent variables)
s = number of observed variables (12)

$$24 < 78$$

Based on the above result, it is confirmed that the model (Figure 9) is over identified, meaning that we have more than enough observed variables (12) to identify the unobserved items (24).

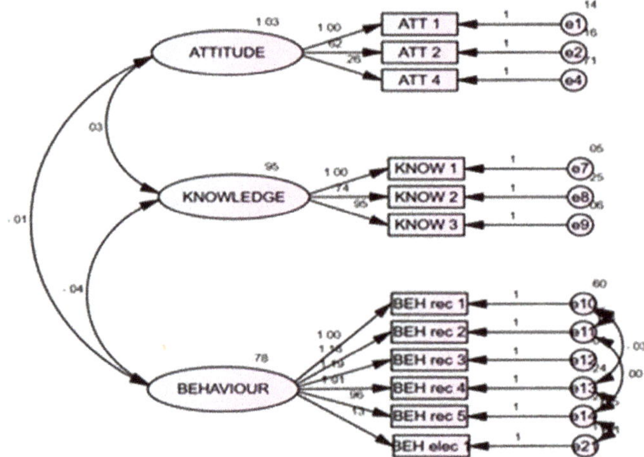

Figure 9. Best fitting measurement model of interrelations among building occupants' environmental AKB after Step 3 (Final Measurement Model).

The list below checks all the parameters in Table 4 related to the data for Figure 9:

- 4 CMIN/DF is 1.273 which is a good fit.
- GFI is 0.954, which ranges between 0 and 1 and is completely satisfactory.
- AGFI is 0.922 and it is a good fit.
- IFI is 0.994 is within the range and closer to 1, which is a good fit.
- NFI is 0.974 and it is satisfactory.
- TLI is 0.992, which is within the range and is the best fit and satisfactory.
- CFI is 0.994, which is within the range and is a good fit.
- RFI is 0.962 and it is a good fit.
- RMR is 0.044 and it is within the range, while it is a better fit, as it is smaller and closer to 0.
- RMSEA is smaller than 0.08, at 0.037, which is a close fit. In Table 4, LO 90 is 0.000 and HI 90 is 0.063 which constitute the lower limit and upper limit of a 90% confidence interval for the population value of RMSEA
- PRATIO is 0.697, which is smaller now, and indicates an acceptable fit.
- PGFI is 0.563, which is an acceptable fit.
- PNFI is 0.679, and as this is closer to 1, it is more acceptable.
- PCFI is 0.693, and it is satisfactory.
- PCLOSE is 0.777, which is now more than 0.05, RMSEA is 0.037, which is less than 0.05; therefore, the hypothesised model (Figure 9) is confirmed, as it indicates all good fits.

The above list confirms that Figure 9 is a good model fit after three steps of modifications on the model. This suggests the finding that many of the observed variables (questionnaire results) did not measure their latent variable 'Behaviour'. Therefore, occupants did not behave in an environemtally friendly manner in all of their daily actions, despite having environmental knowledge, beliefs, and attitudes.

Table 4. Best fitting measurement model

CMIN	CMIN 58.560	DF 46	P 0.101	CMIN/DF 1.273	
RMR, GFI	RMR 0.044	GFI 0.954	AGFI 0.922	PGFI 0.563	
Baseline Comparisons	NFI 0.974	RFI 0.962	IFI 0.994	TLI 0.992	CFI 0.994
Parsimony-Adjusted Measures	PRATIO 0.697	PNFI 0.679	PCFI 0.693		
RMSEA	RMSEA 0.037	LO 90 0.000	HI 90 0.063	PCLOSE 0.777	

4.4. The Structural Model

The interrelationships among unobserved/latent variables and their observed/measured variables were analysed to reach the best fitting measurement model (Figure 9, Table 4). At this stage, where we are confident that the observed variables support their latent variables, we are now ready to create the structural model to identify and verify the interrelationships between latent/unobserved variables. For the latent variables, the direction of the arrows from Attitude and Knowledge toward Behaviour and the correlating arrows between Knowledge and Attitude were determined based on the literature review [5,10,20,60]. The observed variables remain the same as in Figure 9, as their interrelationship with their latent variables are confirmed in Table 4.

The structural model is a hypothesized/conceptualised model that defines interrelationships among the latent (unobserved) variables only, and describes how particular latent variables directly or indirectly influence other latent variables in the model [58]. The resultant structural model is shown in Figure 10.

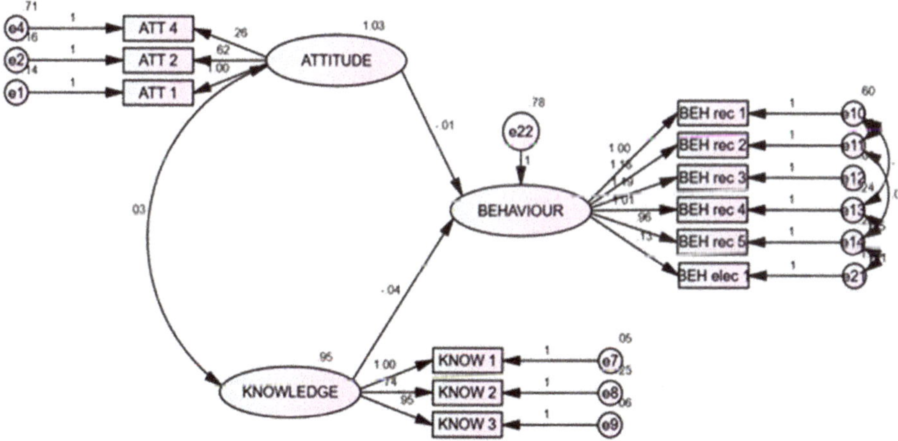

Figure 10. Best fitting structural model.

In Table 5, which is the result for Figure 10, the estimate for regression weight, covariance and correlation are all below 0.40, without a significant p-value (the p-value indicates whether an observation results due to random occurrences or occurs due to a change that was made). This is because estimate values equal to or greater than 0.40 with a significant p-value < 0.05 indicate strong measurement, while values closer to 1 indicate a stronger measurement [42]. A smaller p-value $<$ 0.05 is at the level of significance and indicates that there is stronger evidence, assuming that the null hypothesis is true. The only estimated values with significant p-values were those for Knowledge variance at 0.952, and that for ATTITUDE variance at 1.033. Therefore, the interrelationships were not very significant among these three latent variables.

Table 5. Estimates/scalar estimates/maximum likelihood estimates.

			Estimate	S.E.	C.R.	P
Regression Weights			Estimate	S.E.	C.R.	P
	BEHAVIOUR <-	KNOWLEDGE	−0.038	0.065	−0.587	0.557
	BEHAVIOUR <-	ATTITUDE	−0.006	0.065	−0.086	0.931
Covariances			Estimate	S.E.	C.R.	P
	ATTITUDE <->	KNOWLEDGE	0.028	0.074	0.379	0.705
Correlations			Estimate			
	ATTITUDE <->	KNOWLEDGE	0.028			
Variances			Estimate	S.E.	C.R.	P
	ATTITUDE		1.033	0.205	5.030	***
	KNOWLEDGE		0.952	0.101	9.406	***

*** *p*-value is significant.

The standard errors (S.E.) do not present any extremely large or small values (outliers), and as suggested by Byrne (2010) [42], the model is a reasonably good fit; on the other hand, (S.E.) should not be an extremely small value close to zero, as this indicates a poor model based on the explanation by Bentler et al. (1980) [42,55]. The critical ratios (CR) present small and/or negative values for regression weight, covariance, and correlation, and were only high for KNOWLEDGE and ATTITUDE variance; therefore, no correlations between the latent variables were totally supported. In other words, there are no strong interrelationships among AKB in this research, as the figures on the arrows between the latent variables would need to be closer to 0.1; however, each of them is supported by its observed variables and GOF indices in Table 6 (GOF measures), which is discussed in Section 5 of this paper.

Table 6. Summary of GOF measurement results: (i) conceptual (ii) best fitting measurement and structural models.

GOF Measures	GOF Measures (Recommended)	Conceptual Measurement Model	Best Fitting Measurement & Structural Model
CMIN/DF	1 (very good)–2 (threshold)	6.187	1.273
RMSEA	>0.05 (very good)–0.1 (threshold)	0.160	0.037
RMR	0–1 (Smaller values = better fit)	0.133	0.044
GFI	0 (no fit)–1 (perfect fit)	0.478	0.954
CFI	0 (no fit)–1 (perfect fit)	0.412	0.994
IFI	0 (no fit)–1 (perfect fit)	0.417	0.994
TLI	0 (no fit)–1 (perfect fit)	0.372	0.992

5. Discussion

The results from the survey analysis revealed that the occupants were aware of climate change, and that the majority believed that it was either a major or minor threat. Most of them also confirmed that their high energy consumption could have negative impact on the climate change. However, these attitudes and beliefs were not transformed into environmentally friendly behaviour. For example, when they were asked if they used their washing machine economically, the majority answered occasionally or never. This demonstrated a lack of interrelationship between occupant Attitude, Knowledge and Behaviour (AKB).

To explore this further, an in-depth investigation of the interrelationships among different latent/unobserved variables (AKB) was carried out based on their indicators/observed variables (survey questionnaire results) using SEM.

After several iterations, the model depicted in Figure 10 was the best conceptualised structural model. Even though the structural model was designed based on other existing research findings, interrelationships with a statistical significance level of $p < 0.05$ were rejected in this research.

Table 6 compares the GOF measurements for the three models for: (i) conceptual measurement model, (ii) the best fitting measurement and structural model. It can be seen that there is a remarkable improvement in the outcome of the best fit measurement model compared to the conceptual model, where the GOF does not change for the best fitting structural model, as there are no further changes in the observed variables from the best fitting measurement model. Even though the structural model was a good fit, the interrelationships between AKB were not strong, and therefore, we further reviewed the factor loading between AKB (Table 5). The key difference between Figures 9 and 10 is:

- Behaviour does not affect Attitude or Knowledge, while in Figure 9, they are all interrelated and affect each other, as it was a measurement model.
- The authors made one-way relationships from Attitude and Knowledge to Behaviour; therefore, Behaviour is endogenous. As shown by the arrows in Figure 10, Attitude to Behaviour has a value of −0.01, and Knowledge to Behaviour has a value of −0.04. Based on this data, their interrelationship is not significant.

These results indicate a lack of statistically significant interrelationships among the latent/unobserved variables AKB; however, good model fit was achieved. The insignificant interrelationships among the latent/unobserved variables can viewed as people with good knowledge not necessarily behaving in an environmentally friendly manner; although there are interrelationships among their AKB (confirmed as good model fit in Table 6), the factor loading in Figure 10 among these three latent variables AKB is not close to 0.1. On the other hand, in Table 5, the estimates for regression weight, covariance and correlation are all below 0.40, with insignificant p-values.

Another important finding from this part of the research analysis and results revealed that appropriate attitudes and certain beliefs among the occupants did not necessarily influence their environmental behaviour. Although Figures 9 and 10 are good fit measurements and structural models, respectively, these models were achieved after eliminating some of the observed variables (energy-related behaviours) in order to achieve good model fit in step 1 of the modification process. This was based on the comments from the questionnaire, which indicated that although the majority of the respondents were aware of the need for energy conservation due to the threat of climate change, they still behaved in an un-environmentally friendly manner, e.g., by running the washing machine on a setting higher than on eco-mode, or by leaving the tap on while brushing their teeth. This implies that the environmental behaviour of occupants is not in line with their knowledge and attitudes, as they are not careful with their water and electricity consumption. This uncovers the scope to further identify effective motivational factors in order to reinforce the interrelationships among unobserved variables in order to develop a model that could influence occupant behaviour. One of the strongest motivational factors could be a cash incentive; however, this factor might not be the most powerful motivational factor for occupants in Dubai, due to the salary-to-energy bill ratio, which is affordable for its residents. Future work needs to incorporate reinforcing factors in the best fitting structural model (Figure 10) designed in this study to strengthen AKB interrelationships with the aim of improving the environmental behaviour of building occupants.

6. Conclusions

Increasing energy consumption and the resultant carbon emissions in residential buildings is becoming a critical issue that should be focused upon to achieve a green built environment and to mitigate global warming. Green rating systems such as the Leadership in Energy and Environmental Design (LEED) and many others are being actively practised in various regions of the world that aspire to achieve energy savings, reductions in carbon emissions, and occupant satisfaction. There is however, still a considerable performance gap which exists between energy consumption as-designed and the one in the actual buildings. Occupant behaviour accounts for one of the major reasons behind the significant uncertainty regarding building energy use. Little is known about how the occupants of these buildings cause the performance gap.

Green building initiatives, such as LEED and many other good practices, are now regulated by the UAE government in an attempt to achieve energy conservation, sustainability, health, and environmental and economic benefits. The key research question was whether the occupants of these buildings have the knowledge required to change and/or improve their environmental behaviour in order to achieve energy savings, and whether their beliefs and attitudes could lead them towards greener behaviour. The main novelties and objectives of the present work can be summarized as:

Data was collected from four LEED-certified multi-residential buildings in Dubai, UAE to better understand occupant behaviour and their level of involvement in building operations. A total of 203 occupants responded to the survey with valid answers, resulting in a response rate of 76.6%. Survey data was initially tested using SPSS statistics software to obtain descriptive statistics, frequencies, and means. Structural Equation Modelling (SEM) was then used to perform an in-depth analysis of the interrelationships between the three unobserved/latent variables—occupant environmental Attitude, Knowledge and Behaviour (AKB)—based on the observed/measured variables (survey data).

The comparison between the conceptual measurement model and the best fit measurement and structural models revealed significant improvement in all parameters. The final parameters fell within the GOF range, i.e., CMIN/DF improved from 6.187 to 1.273, RESEA from 0.160 to 0.037, RMR from 0.133 to 0.044, GFI from 0.478 to 0.954, CFI from 0.412 to 0.994, IFI from 0.417 to 0.994, and finally TLI from 0.372 to 0.992. On the other hand, the final regression weight from Knowledge to Behaviour was 0.557, and that from Attitude to Behaviour was 0.931; in both cases, the p-value was not less than 0.05, and therefore no significant interrelationship exists between them.

For the best fitting structural model, the estimates for regression weights, covariances and correlations were all below 0.40, without a significant p-value. The standard errors (S.E.) at 0.065 indicated that the model provided a reasonably good fit, as it was between 0 and 1, although being closer to zero indicates a poorer model, based on the explanation by Bentler et al. (1980) [42,55]. The critical ratios (CR) were either small or negative values, e.g., Attitude to Behaviour was -0.086, and the value for covariances among Attitude and Knowledge was 0.379. The CR was only high for variances for Knowledge, at 5.030, and for Attitude, at 9.406, thus indicating none of the correlations between the latent variables were totally supported.

These results indicate a lack of statistically significant interrelationships between the latent variables; however, a significant relationship exists between the latent and the measured/observed variables. In other words, occupants with good attitudes do not necessarily behave in an environmentally friendly manner. Well-fit measurements and structural models were obtained, but only after the elimination of observed variables related to electricity and water consumption in Step 1 of the modification (as explained in Section 5). This reflects the fact that occupant environmental behaviour is not in line with design intent and confirms that "*the occupants of the LEED-certified buildings in UAE are not motivated to behave in an environmentally friendly manner*", by showing the insignificant relationship between the latent variables, AKB.

Since there is no apparent attention being paid to, or concern being expressed with respect to, occupant behaviour in the post-occupancy phase, it is suggested that the LEED-certification process should include a recommissioning/recertification phase after occupancy that includes occupant behaviour measurements and revision. Some of the real potential applications of the developed methodology, for instance, could be a public organisation and/or governmental authority in charge of energy efficiency in the built environment to review and monitor the interrelationships among AKB in the building's occupants to provide revisions to LEED, rather than changing the LEED process, which is intended to value the work carried out during construction and design of the buildings.

The research findings present a logical guide and a well-fitting structural model (Figure 10), which can form the basis for the development of a Building Occupant Environmental Behaviour model through the inclusion of motivational factors to trigger environmentally friendly actions and influence behaviour. It is accepted that today it is quite possible to produce an eco-friendly green building; however, it is also a priority need to adopt practices such as Soft Landings [63,64] in order to involve industry professionals and educate occupants with the aim of improving and altering their behaviour in an environmentally friendly fashion to achieve potential energy savings in their buildings.

Author Contributions: Conceptualization, E.N. and M.S.G.; methodology, E.N.; software, E.N.; validation, E.N. and M.S.G.; formal analysis, E.N.; investigation, E.N. and M.S.G.; resources, E.N. and M.S.G.; data curation, E.N. and M.S.G.; writing—original draft preparation, E.N. and M.S.G.; writing—review and editing, M.S.G.; visualization, M.S.G.; supervision, M.S.G.; project administration, M.S.G.; funding acquisition, E.N. All authors have read and agreed to the published version of the manuscript.

Funding: This research received no external funding.

Acknowledgments: The authors would like to thank the survey participants to carry out this research

Conflicts of Interest: The authors declare no conflict of interest.

List of Abbreviations

AGFI	Adjusted Goodness of Fit Index
AKB	Attitude, Knowledge and Behaviour
AMOS	Analysis of Moment Structure
BEPAC	Building Environmental Performance Assessment Criteria
BREEAM	Building Research Establishment Environmental Assessment Methodology
CASBEE	Comprehensive Assessment System for Built Environment Efficiency
CFA	Confirmatory Factor Analysis
CFI	Comparative Fit Index
CMIN/DF	Ratio of Minimum Discrepancy to the Degree of Freedom
CO2	Carbon Dioxide
CR	Critical Ratios
GBC	Green Building Council
FSEC	Florida Solar Energy Center
GFI	Goodness of Fit Index
IFI	Incremental Fit Index
LEED	Leadership in Energy and Environmental Design
MI	Modification Indices
NFI	Normed Fit Index
NGO	Non-Government Organization
NNFI	Non-Normed Fit Index
PCFI	Parsimony Comparative Fit Index
PCLOSE	P of CLOSE fit for RMSEA
PGFI	Parsimony Goodness of Fit Index
PMV	Predicted Mean Vote
PNFI	Parsimony Normed Fit Index
POE	Post Occupancy Evaluation
PPD	Percentage of People Dissatisfied
PRATIO	Parsimony RATIO
P-VALUE	Probability of observing a result
RFI	Relative Fit Index
RMR	Root Mean square Residual
RMSEA	Root Mean Square Error of Approximation
SE	Standard Error
SEM	Structural Equation Modelling
SMUD	Sacramento Municipal Utility District
TLI	Tucker-Lewis Index
UAE	United Arab Emirates

References

1. Ji, W.Y.; Chan, E.H.W. Critical factors influencing the adoption of smart home energy technology in China: A Guangdong province case study. *Energies* **2019**, *12*, 4180. [CrossRef]
2. Papadopoulos, A.M.; Giama, E. Rating systems for counting buildings environmental performance. *Int. J. Sustain. Energy* **2009**, *28*, 29–42. [CrossRef]
3. Ghaffarianhoseini, A.; Dahlan, N.D.; Berardi, U.; Ghaffarianhoseini, A.; Makaremi, N.; Ghaffarianhoseini, M. Sustainable energy performances of green buildings: A review of current theories, implementations and challenges. *Renew. Sustain. Energy Rev.* **2013**, *25*, 1–17. [CrossRef]
4. Birt, B.; Newsham, G.R. Post-occupancy evaluation of energy and indoor environment quality in green buildings: A review. In Proceedings of the 3rd International Conference on Smart and Sustainable Built Environments, Delft, The Netherlands, 15–19 June 2011; pp. 15–19.
5. D'Oca, S.; Hong, T.; Langevin, J. The human dimensions of energy use in buildings: A review. *Renew. Sustain. Energy Rev.* **2018**, *81*, 731–742. [CrossRef]
6. Turner, C.; Frankel, M. *Energy Performance of LEED for New Construction Buildings—Final Report*; New Buildings Institute: White Salmon, WA, USA, 2008.
7. Poirazis, H.; Blomsterberg, M. Wall Energy simulations for glazed office buildings in Sweden. *Energy Build.* **2008**, *40*, 1161–1170. [CrossRef]
8. Socolow, R.; Sonderegger, R. *The Twin Rivers Program on Energy Conservation in Housing: Four-Year Summary Report*; Centre for Environmental Studies: Princeton, NJ, USA, 1976.
9. Nguyen, T.A.; Aiello, M. Energy intelligent buildings based on user activity: A survey. *Energy Build.* **2013**, *56*, 244–257. [CrossRef]
10. Schipper, L.; Bartlett, S.; Vine, E. Linking lifestyle and energy use: A matter of time? *Annu. Rev. Energy* **1989**, *14*, 273–320. [CrossRef]
11. Parker, D.; Mazzara, M.; Sherwin, J. Monitored energy use pattern in low-income housing in a hot and humid climate. In Proceedings of the Tenth Symposium on Improving Building Systems in Hot Humid Climates, Fort Worth, TX, USA, 13–14 May 1996; Volume 316.
12. SWA. *Cutting the Power, CARB News, Consortium for Advanced Residential Buildings*; Stephen Winter and Associates, US Department of Energy: New York, NY, USA, 2005; Volume 8, p. 9.
13. Gram-Hanssen, K.; Georg, S.; Christiansen, E.T.; Heiselberg, P.K. How building regulations ignore the use of buildings, what that means for energy consumption and what to do about it. In Proceedings of the Summer Study, ECEEE, European Council for an Energy Efficient Economy, Hyeres, France, 29 May–3 June 2017.
14. Gram-Hanssen, K.; Georg, S. Energy performance gaps; promises, people, practices. *Build. Res. Inf.* **2018**, *46*, 1–9. [CrossRef]
15. Kim, S.; Song, Y.; Sung, Y.; Seo, D. Development of a consecutive occupancy estimation framework for improving the energy demand prediction performance of building energy modeling tools. *Energies* **2019**, *12*, 433. [CrossRef]
16. Gul, M.S.; Patidar, S. Understanding the energy consumption and occupancy of a multi-purpose academic building. *Energy Build.* **2015**, *87*, 155–165. [CrossRef]
17. Congedo, P.M.; Baglivo, C.; Centonze, G. Walls comparative evaluation for the thermal performance improvement of low-rise residential buildings in warm Mediterranean climate. *J. Build. Eng.* **2020**, *28*, 101059. [CrossRef]
18. Zhang, Y.; Bai, X.; Mills, F.P. Characterizing energy-related occupant behavior in residential buildings: Evidence from a survey in Beijing, China. *Energy Build.* **2020**, *214*, 109823, ISSN 0378-7788. [CrossRef]
19. Ashouri, M.; Haghighat, F.; Fung, B.C.M.; Yoshino, H. Development of a ranking procedure for energy performance evaluation of buildings based on occupant behaviour. *Energy Build.* **2019**, *183*, 659–671, ISSN 0378-7788. [CrossRef]
20. Blanke, J.; Beder, C.; Klepal, M. An integrated behavioural model towards evaluating and influencing energy behaviour—The role of motivation in behaviour demand response. *Buildings* **2017**, *7*, 119. [CrossRef]
21. Parys, W.; Saelens, D.; Hens, H. Implementing realistic occupant behaviour in building energy simulations—The effect on the results of an optimization of office buildings. In Proceedings of the 10th REHVA World Congress, Antalya, Turkey, 9–12 May 2010.

22. Belafi, Z.; Hong, T.; Reith, A. A critical review on questionnaire surveys in the field of energy-related occupant behaviour. *Energy Effic.* **2018**, *11*, 2157–2177. [CrossRef]
23. Hamad, F.; Abu-Hijleh, B. The energy savings potential of using dynamic external louvers in an office building. *Energy Build.* **2010**, *42*, 1888–1895. [CrossRef]
24. AboulNaga, M.M.; Elsheshtawy, Y.H. Environmental sustainability assessment of buildings in hot climates: The case of the UAE. *Renew. Energy* **2001**, *24*, 553–563. [CrossRef]
25. Shanks, K.; NezamiFar, E. Impact of climate change on building cooling demands in the UAE. In Proceedings of the SB13 Dubai: Advancing the Green Agenda Technology, Practices and Policies, Dubai, UAE, 8–12 December 2013.
26. Radhi, H.; Fikiry, F. A Statistical model to assess Indirect CO_2 Emissions of the UAE residential sector. In Proceedings of the Tenth international conference for enhanced building operations, Kuwait City, Kuwait, 26–28 October 2010. ESL-IC-10-10-60.
27. Salama, M.; Hana, A.R. Green buildings and sustainable construction in the United Arab Emirates, UAE. In Proceedings of the 26th Annual ARCOM Conference, Leeds, UK, 6–8 September 2010; pp. 1397–1405.
28. Ward, A. Positive "Green" Approach Urged. 2009. Available online: http://www.cmguide.org/archives/1524 (accessed on 15 June 2013).
29. Landais, E. Dubai to Turn Green in 2008. 2007. Available online: http://m.gulfnews.com/dubai-to-turn-green-in-2008-1.207817?utm_referrer (accessed on 6 November 2019).
30. Maceda, C. UAE Leads the Region in Sustainability Trend, Gulf News. 2013. Available online: http://gulfnews.com/business/general/uae-leads-region-in-sustainability-trend-1.1199896 (accessed on 15 June 2013).
31. The United Arab Emirates (UAE) Cabinet, Highlights of the UAE Government Strategy. 2011. Available online: https://u.ae/en/about-the-uae/the-uae-government/the-uae-cabinet (accessed on 15 June 2013).
32. Expo2020-Dubai. Available online: https://www.expo2020dubai.com/en/discover/sustainability (accessed on 6 November 2019).
33. Masdar City, New Energy Economy, Past and Present Companies. 2013. Available online: http://www.masdar.ae/en/home/detail/past-and-present-companies (accessed on 15 June 2013).
34. Gulf News, the UAE, Dubai Faces Sustainability Challenges: Report. 2017. Available online: https://gulfnews.com/uae/environment/dubai-faces-sustainability-challenges-report-1.2102107 (accessed on 6 November 2017).
35. Reposa, J.H. Comparison of USGBC LEED for Homes and the NAHB National Green Building Program. *Int. J. Constr. Educ. Res.* **2009**, *5*, 108–119. [CrossRef]
36. Behbehani, L.J. Does LEED-Certified Multifamily Housing Influence Occupants Environmental Behaviors? Case Studies of Residential Development in the Midwest. Ph.D. Thesis, Purdue University, West Lafayette, IN, USA, 2012.
37. Jones, P.; Vyas, U.K. Energy Performance in Residential Green Developments: A Florida Case Study. *Real Estate Issues* **2008**, *33*, 65–71.
38. Wood, G.; Newborough, M. Dynamic energy-consumption indicators for domestic appliances: Environment, behaviour and design. *Energy Build.* **2003**, *35*, 821–841. [CrossRef]
39. Vlek, C.; Steg, L. Human Behavior and Environmental Sustainability: Problems, Driving Forces, and Research Topics. *J. Soc. Issues* **2007**, *63*, 1–19. [CrossRef]
40. Riley, M.; Kokkarinen, N.; Pitt, M. Assessing post occupancy evaluation in higher education facilities. *J. Facil. Manag.* **2010**, *8*, 189–220. [CrossRef]
41. USGBC. *LEED for New Construction*; USGBC: Washington, DC, USA, 2014.
42. Byrne, B.M. *Structural Equation Modelling with Amos: Basic Concepts, Applications, and Programming*, 2nd ed.; Rutledge, Taylor and Francis Group: New York, NY, USA, 2010.
43. Hair, J.F.; Ringle, C.M.; Sarstedt, M. Partial least squares: The better approach to structural equation modeling? *Long Range Plan* **2012**, *45*, 312–319. [CrossRef]
44. Hair, J.F.; Ringle, C.M.; Sarstedt, M. Partial least squares structural equation modeling: Rigorous applications, better results and higher acceptance. *Long Range Plan* **2013**, *46*, 1–12. [CrossRef]
45. Lowry, P.B.; Gaskin, J. Partial least squares (PLS) structural equation modeling (SEM) for building and testing behavioural casual theory: When to choose it and how to use it. *IEEE Trans. Dependable Secur. Comput.* **2014**, *57*, 123–146.

46. Akter, S.; Wamba, S.F.; Dewan, S. Why PLS-SEM is suitable for complex modelling? An empirical illustration in big data analytics quality. *Prod. Plan. Control* **2017**, *28*, 1011–1021. [CrossRef]
47. Hulland, J. Use of partial least squares (PLS) in strategic management research: A review of four recent Studies. *Strat. Manag. J.* **1999**, *20*, 195–204. [CrossRef]
48. Hair, J.; Hollingsworth, C.L.; Randolph, A.B.; Chong, A.Y.L. An updated and expanded assessment of PLS-SEM in information systems research. *Ind. Manag. Data Syst.* **2017**, *117*, 442–458. [CrossRef]
49. Rigdon, E.E. Choosing PLS path modeling as analytical method in European management research: A realist perspective. *Eur. Manag. J.* **2016**, *34*, 598–605. [CrossRef]
50. Sarstedt, M.; Ringle, C.M.; Smith, D.; Reams, R.; Hair, J.F. Partial least squares structural equation modeling (PLS-SEM): A useful tool for family business researchers. *J. Fam. Bus. Strat.* **2014**, *5*, 105–115. [CrossRef]
51. Usakli, A.; Kucukergin, K.G. Using partial least squares structural equation modeling in hospitality and tourism: Do researchers follow practical guidelines? *Int. J. Contemp. Hosp. Manag.* **2018**, *30*, 3462–3512. [CrossRef]
52. Lee, L.; Petter, S.; Fayard, D.; Robinson, S. On the use of partial least squares path modeling in accounting research. *Int. J. Acc. Inf. Syst.* **2011**, *12*, 305–328. [CrossRef]
53. Aibinu, A.A.; Al-Lawati, A.M. Using PLS-SEM technique to model construction organizations' willingness to participate in e-bidding. *Autom. Constr.* **2010**, *19*, 714–724. [CrossRef]
54. Hair, J.F.; Sarstedt, M.; Ringle, C.M.; Mena, J.A. An assessment of the use of partial least squares structural equation modeling in marketing research. *J. Acad. Mark. Sci.* **2012**, *40*, 414–433. [CrossRef]
55. Bentler, P.M.; Bonnet, D.C. Significance of tests and goodness of fit in the analysis of covariance structures. *Psychol. Bull.* **1980**, *88*, 580–606. [CrossRef]
56. Islam, M.D.M.; Faniran, O.O. Structural equation model of project planning effectiveness. *Constr. Manag. Econ.* **2005**, *23*, 215–223. [CrossRef]
57. Hair, J.F.; Anderson, R.E.; Tatham, R.L.; Black, W.C. *Multivariate Data Analysis*, 5th ed.; Prentice Hall: Upper Saddle River, NJ, USA, 1998.
58. Byrne, B.M. *Structural Equation Modelling with Amos: Basic Concepts, Applications and Programming*, 3rd ed.; Rutledge, Taylor and Francis Group: New York, NY, USA, 2016.
59. Bodoff, D.; Ho, S.Y. Partial Least Squares Structural Equation Modeling Approach for Analysing a Model with a Binary Indicator as an Endogenous Variable. *Commun. Assoc. Inf. Syst.* **2016**, *38*, 400–419.
60. Van Raaij, W.F.; Verhallen, T.M.M. A behavioral model of residential energy use. *J. Econ. Psychol.* **1983**, *3*, 39–63. [CrossRef]
61. Iacobucci, D. Structural equations modelling: Fit indices, sample size, and advanced topics. *J. Consum. Psychol.* **2010**, *20*, 90–98. [CrossRef]
62. Kline, R.B. *Principles and Practice of Structural Equation Modelling: Methodology in the Social Sciences*; Guilford Press: New York, NY, USA, 2005.
63. BSRIA. Soft Landings. 2019. Available online: https://www.bsria.com/uk/consultancy/project-improvement/soft-landings/ (accessed on 3 May 2019).
64. Bordass, B. Soft landings: Can We Make Follow-Through, Feedback and POE Routine for Design and Building Teams, Presented to Lawrence Berkley National Laboratory (LBL). 2008. Available online: http://www.usablebuildings.co.uk (accessed on 3 May 2019).

Article

Feasibility Study of Integrating Renewable Energy Generation System in Sark Island to Reduce Energy Generation Cost and CO_2 Emissions

Shamir Robinson [1], Savvas Papadopoulos [1], Eulalia Jadraque Gago [2,*] and Tariq Muneer [1]

1 School of Engineering and the Built Environment, Edinburgh Napier University, Edinburgh EH10 5DT, UK; Shamirrobinsonlora@gmail.com (S.R.); S.Papadopoulos@napier.ac.uk (S.P.); T.Muneer@napier.ac.uk (T.M.)

2 School of Civil Engineering, University of Granada, 18071 Granada, Spain

* Correspondence: ejadraque@ugr.es

Received: 15 October 2019; Accepted: 3 December 2019; Published: 11 December 2019

Abstract: The island of Sark, located in the English Channel, has endured an electricity distribution crisis for the past few years, resulting in high electricity costs almost six times higher than UK mainland energy prices. This article is focused on a methodology for finding the best renewable energy system with the lowest levelized cost of energy (LCOE) in comparison to the current energy rate of 66 p/kWh. Three different main cases of study have been compared in performance for different levels of renewable energy integration and energy storage, evaluating the estimated size of the system, installation cost and CO_2 emissions. The results, which depend on the assumptions outlined, show that Case 2 renewable energy generation system is the most suitable in terms of reduction of CO_2 emissions and expected earnings from a lower LCOE. Uncertainty in the results could be minimized if actual data from the island is made available by following the same methodology to find the best solution to the island's current energy generation problem. Due to non-available data for the load profiles and wind velocity a set of assumption were required to be implemented. As such, two different load profiles were selected—one with a peak of energy consumption in winter and the other with a summer peak.

Keywords: photovoltaic; wind energy; storage; islanded systems

1. Introduction

There is a global challenge to reduce the greenhouse gas (GHG) emission to the environment by 2050. For this reason, the UK has committed to decrease GHG emissions by 80% compared to 1990 levels [1] and to achieve net zero carbon emissions by 2050. Two main trends to reduce emissions are to either reduce the carbon intensity of electricity generation by introducing more renewable or low-carbon generation systems, and the other is to reduce average and peak consumption, usually achieved by enhancing the efficiency of the overall distribution system and by influencing consumption patterns.

Due to this challenge, there has been an accelerated increase of renewable energy installation in the past 10 years, boosted by a variety of factors like governmental incentives, a drastic drop in manufacturing prices, and technological maturity that helps drive the price of projects lower. With all these advantages, the cost of renewable energies can help reduce overall generating cost.

The growth of renewable energy generation needs to be monitored because a high penetration of renewable energy sources into the grid can create instability resulting in issues with the frequency control and response issues. Furthermore, the intermittent nature of renewables like solar and wind directly affects the grid stability. This effect is even more severe on small islanded systems without interconnection to other generation or loads that can absorb the energy generation surplus.

Energy storage helps to balance the grid by reshaping supply and demand patterns by storing the energy and enabling its use at a later time. There are many different types of energy storage methods, such as mechanical (hydroelectric, pumped-storage, compressed air, etc.), thermal (sensible heat, latent heat, etc.), and electrochemical.

The scope of this study is to validate the feasibility of introducing renewable energy penetration to cover the island's current and near future energy needs, comparing a variety of parameters concerning the installation and sizing. The outcomes of the followed methodology are validated by evaluating key control parameters for each individual scenario considered. Those control values are the installation cost, CO_2 emissions, levelized cost of energy, and the return of investment [2].

This article is focused on finding the best renewable energy system with the lowest LCOE in comparison to the current energy cost of 66 p/kWh [3]. Based on educated assumptions in addition to data analysis and manipulation, we estimate an optimal integration of wind and solar as primary renewable energies with an energy storage system (if required) to stabilize the grid.

Furthermore, besides a high level of performance, the study compares the financial viability of the generation system to obtain a LCOE lower than the actual cost the energy of 66 p/kWh. This article also relates to the goal of the Sark Island energy commissioner on the price control order aiming to achieve an energy cost reduction to 56 p/kWh for the island.

The scope for this article is limited to

- Estimate the island energy profile and different generation systems based on available data and system comparison;
- Use a combination of wind, solar, and battery storage with the current energy system and calculate the technical and financial feasibility assuming a 20-year lifespan;
- Showcase the advantages and disadvantages for three different case studies with different installation costs and operational implications; and
- Use a simplified levelized cost of energy (LCOE) estimation to compare the feasibility of the system.

The main advances of this research, related to different studies conducted in this area, were

(1) This study used all of the available data and implemented educated assumptions to estimate data that were not available (load, wind velocity). Data analysis and manipulation were carried out to estimate an optimal integration of wind and solar as primary renewable energies with an energy storage system to stabilize the grid, if required.
(2) The worst-case scenario was used for the load, running two highly different load profiles with peaks in winter and summer. Only the latter is presented here.
(3) This method takes an hourly data analysis of the variation on the state of charge of the battery for the complete year with the help of three energy generation systems to keep the battery charged.
(4) The validation was based on the use of a theoretical wind velocity estimation using the Rayleigh distribution function from the 2013–2019 data sets.
(5) Monthly solar irradiation values available from NASA daily data were used for Sark Island. Those data were decomposed in hourly values by the present team.
(6) Using MATLAB coding, the available data is analyzed in hourly and 15 min intervals to calculate the optimal battery size and auxiliary energy generation require (gas or diesel) to keep the batteries in a good state. Dalton et al. [4] based their research on the calculation of a medium-sized energy generation system by applying all the calculations using Homer software (Hybrid Optimization Models for Energy Resources) power optimization software by NREL (National Renewable Energy Laboratory). In contrast, the method applied in this research are more open to variabilities and adjustment due to the use of our own coding and calculation algorithms to find the most cost-effective system for an entire island.

(7) The sensitivity analysis in this article evaluates more than 40 different scenarios to compare and validate the best energy mix for the island. Moreover, the way that it is implemented enables the coding to be used to calculate the same output for any other new set of data.

2. Background

The island of Sark, located in the English Channel in the Bay of St Malo, is enduring an electricity distribution crisis. The issue relates to high electricity costs of almost six times higher than UK mainland energy costs. Sark electricity, the island's privately owned energy generation entity, has warned that a reduction in the electricity price could result in a permanent shutdown of energy generation.

The island's independent energy commissioner has objected to the high electricity cost of 66 p/kWh [5] and ordered the reduction of the energy price to 52 p/kWh in a Price Control Order [4]. However, this is still much higher than the UK average of 14 p/kWh.

As a response, Sark Electricity claimed that, if the price reduction is applied, they would lose £20,000 a month. The increase of energy prices has been attributed to a lower demand, forcing an increase in cost to be able to cover the capital cost of its operation. The island government is trying to reach an agreement to buy off the company from the Gordon-Brown Brothers.

The price control law also aims to reduce the cost to 52 p/kWh from August 2018 to 2019 and then reduce it to a minimum of 49 p/kWh from 2019 to 2020. With the current untaxed diesel cost of 42 p/l [3], this goal is difficult to obtain. For that reason, an alternate renewable energy generation system is evaluated to partially or completely generate the energy that the island requires.

3. Methodology

Figure 1 shows the step-by-step workflow diagram with the input data that initially is stored and organized differently with different times between measurements. A brief introduction to the structure of the data, the process to clean it and calculate the values will be presented here.

The methodology was divided in the following four sections:

3.1. Data Research, Calculations and Estimations Analysis

The first part has been to acquire the different sets of data for the island, including load (consumption), wind speed, and solar irradiance, in order to calculate the output and the integrations of the different scenarios. The only calculation in this section is for the solar power output for a single solar panel using standardized data in kWh per 10 min measurement intervals.

3.2. Data Processing

The data processing consists of data cleaning from any the missing or out of range/corrupt data values. Then, the solar and load files have been calculated at 10 min interval values to have more accurate results for the sizing of the battery banks.

3.3. Case Studies

To find the best solution for the island's actual energy requirements, a set of three different main case studies are compared in terms of the performance and integration of the renewable energy, the estimated size of the system, installation cost, and the CO_2 emission.

- Case 1: 100% Renewable energy generation system with energy storage. A purely wind energy generation (500 kW) and an oversized battery bank.
- Case 2: Gen-set and renewable energy generation system with energy storage; 500 kW wind turbine with the addition of using the small generator of 375 KVA to charge the battery when renewable energy is insufficient.

- Case 3: 40% Renewable energy penetration. Mix of solar and wind to have a more stable energy output during the day when the peak on the load occurs with a 150 kW wind turbine and 150 kWp solar-PV system.

3.4. Economical and CO_2 Emissions

One of the main benefits of renewable generation installation is reducing the carbon intensity of the grid. For the UK climate, photovoltaic installations have a carbon intensity of 41 gCO_2e/kWh, which is below the UK 2030 target [6]. From a life cycle study on onshore wind turbines, the carbon intensity is around 60 gCO_2e/kWh [7] (Table 1).

Using MATLAB data analysis for each case, the required installed system capacity has been obtained for the annual energy output and used to compare the CO_2 emissions and the economic viability of the designed system (Figure 1).

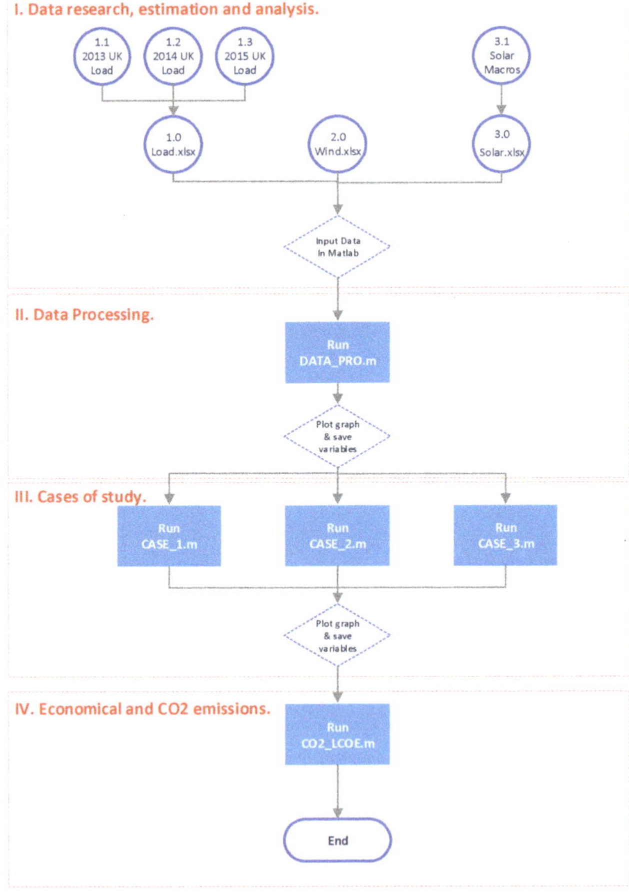

Figure 1. Workflow diagram. Data treatment and MATLAB code processing and results.

Table 1. Renewable energy CO_2 emissions.

	Renewable Energy CO_2 Emissions		
CO_2 Emissions	Solar Photovoltaics panels	41	$kgCO_2/MWh$
	Wind turbine	60	$kgCO_2/MWh$
	Lithium Battery storage	150	$kgCO_2/MWh$
	Diesel Generator	458	$kgCO_2/MWh$

4. Data Research and Processing (Section I and II of the Methodology)

4.1. Load Consumption Estimation

From an official report from the electricity commissioner, the annual total consumption of the island is 1600 mWh/year [3]. Moreover, the hourly consumption profile was chosen to be similar to the United Kingdom as an approximation due to the comparable annual temperature profiles (Figure 2), where the ambient temperature can relate the peak of consumption being in winter [8].

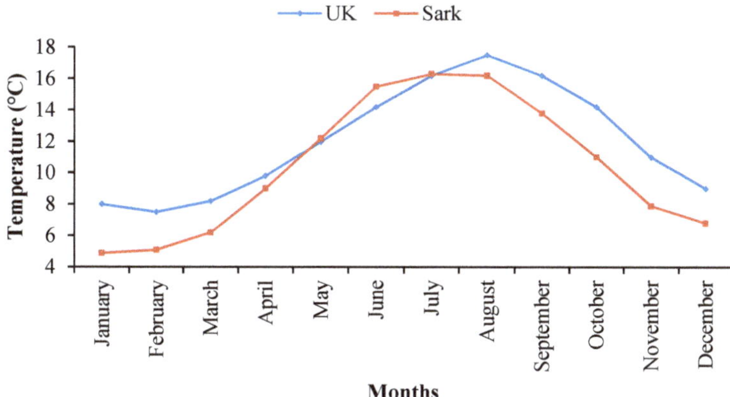

Figure 2. Annual temperature comparison between the UK and Sark Island [9].

Selecting the UK as the characteristic consumption profile for the Sark Island, the first step was to upload the data in MATLAB and clean all the missing point of measures in one-hour intervals. The data was cleaned, and the mean for each hour was calculated between the three years of data available (2013, 2014, and 2015). The data was downloaded from the Entsoe (European Network of Transmission System Operators) [10].

The next step was to adjust the annual UK load (292.70 GWh/Year) to the annual load of the island of 1.6 GWh/year. A correction factor (IS-CF1) has been used to reduce total energy consumed and applied to all the data set (Equation (1)).

$$IS_{CF1} = \frac{\text{Sark Island Annual Consumption}\left(\frac{\text{mWh}}{\text{Year}}\right)}{\text{UK Annual Consumtion}\left(\frac{\text{mWh}}{\text{Year}}\right)} = 0.0055 \quad (1)$$

Figure 3 shows the new adjusted consumption after the correction factor was applied. As a final step in the load analysis, the two most representative days where plotted. The worst day of winter required a load of 286 kW. This step also confirms that the data estimates comply with the actual generator working on Sark Island, with a capacity of 600 kW [11]. On the other hand, this also shows the lowest consumption of 108.6 kW for the summer day with the least generation required.

Using these two values, the highest generation requirement for the system of battery storage and the minimum energy that needs to be supply at all times can be determined (Figure 4).

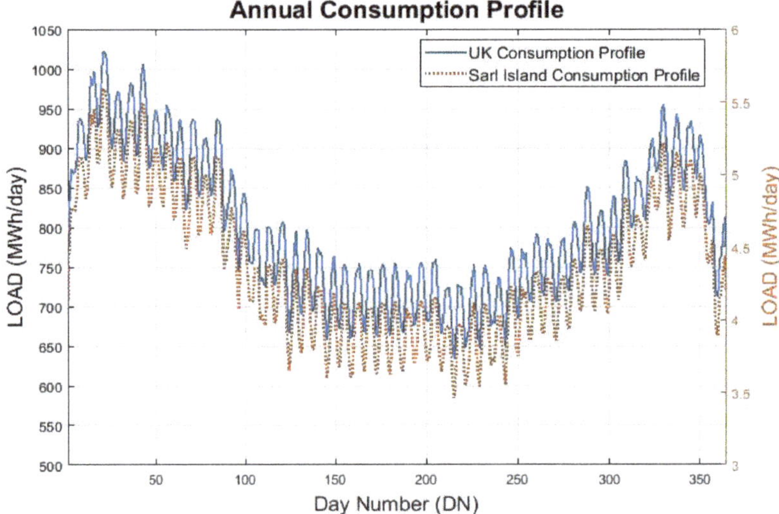

Figure 3. Annual energy consumption for the UK [10].

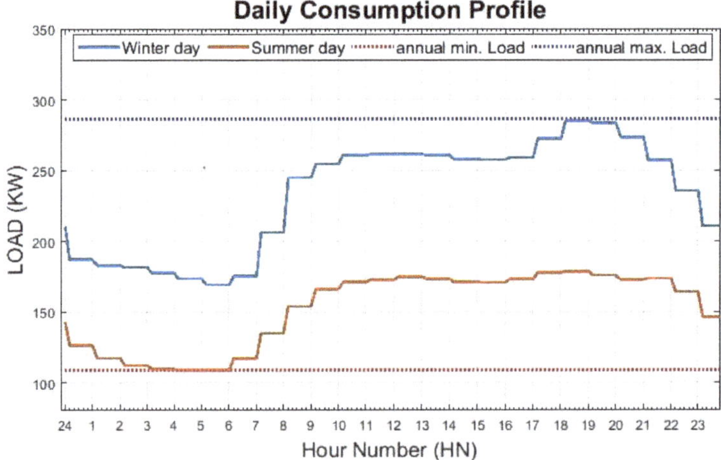

Figure 4. Daily energy consumption profile from Sark Island.

4.2. Wind Generation Estimation

One of the biggest challenges in this research was access to accurate data. To be able to get wind data in 10 minutes interval measurement, the monthly data available has been compared using data from NASA [9] for the location of Sark Island latitude and longitude (49.432–2.36) and a set of available data from North Mains of Cononsyth farm located in Arbroath, Aberdeenshire (56.30–2.44) in the required measurement intervals. Figure 5 shows the close monthly mean between the two locations with an average 10% deviation of the wind speed.

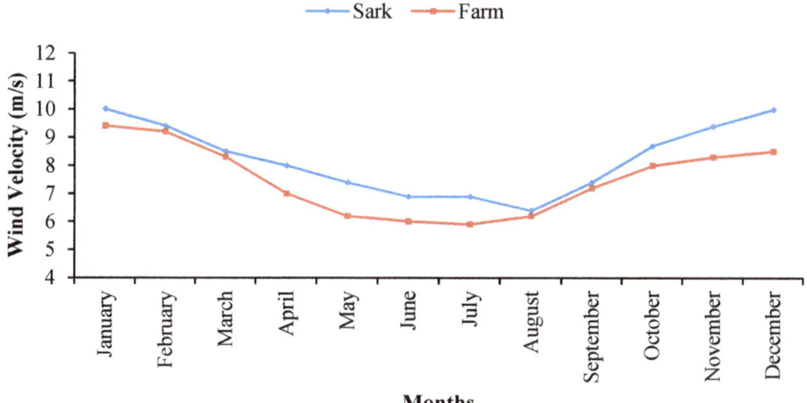

Figure 5. Mean wind velocity at 10 m.

4.2.1. Cononsyth Farm Wind Data Analysis

From available data of 10 minutes interval reading from a weather station installed on the Cononsyth Farm premise, we produced the following study for the wind behaviors and the direct output of the installed 330 Enercon wind turbine. To validate the use of a theorical wind velocity estimation, a Rayleigh distribution function ((Equation (2)) was applied to the data set from 2013 to 2019 (Figure 6) and then compared to the theorical one (Figure 7).

$$S = \hat{U}\left(\frac{2}{\pi}\right)^{0.5}; P(U) = \frac{U}{S^2}\exp\left(-\frac{U^2}{2S^2}\right), \tag{2}$$

where

U = wind speed;
Û = annual mean wind speed.

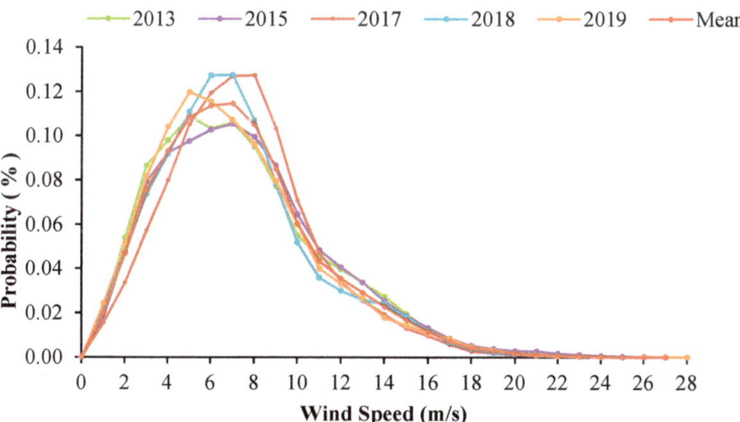

Figure 6. Rayleigh distribution function (2013–2019).

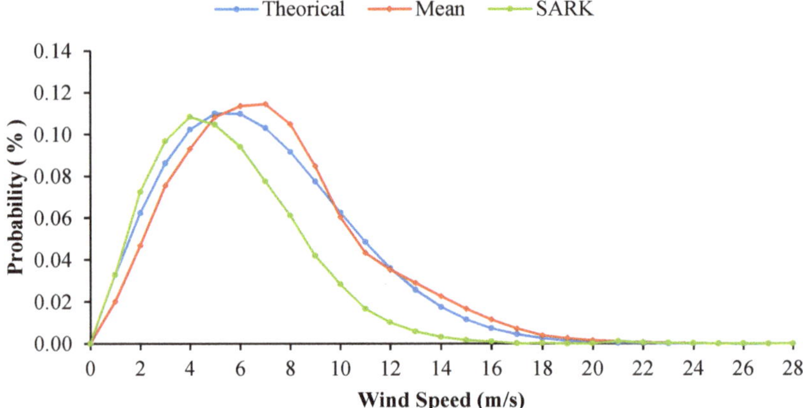

Figure 7. Comparison between Rayleigh distribution function and theorical function.

It is shown that there is a close relation between the theorical curve and the actual wind speed behavior on the island. Compared with the Sark Island distribution analysis, a variance of the peak wind speed can be appreciated. For Sark, the mean is 4.2 m/s, versus a mean of 6.4 m/s for Cononsyth Farm.

4.2.2. Wind Turbine Power Output Calculation

As a demonstrative year, Figure 8 shows the power output of the 330 kW wind turbine for the current 2019. Figure 9 shows the data sets for 2013, 2015 and 2018 output were also analyzed and included in the mean.

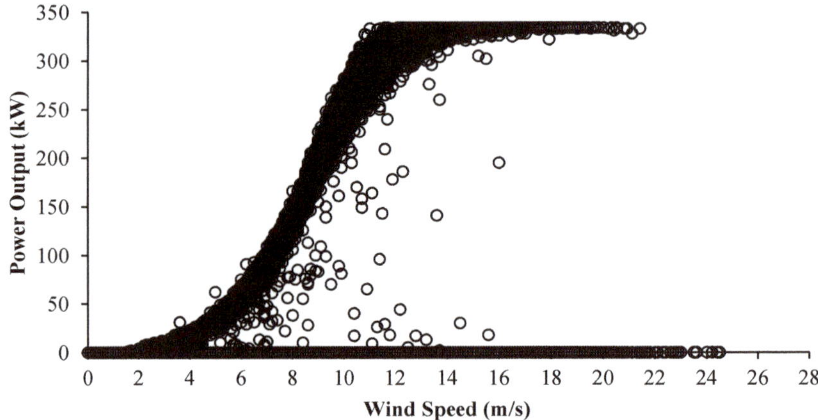

Figure 8. Power output for Enercon 330 kW (2019).

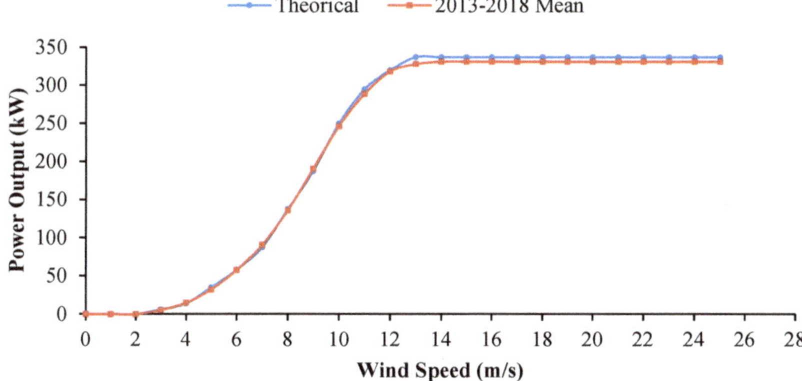

Figure 9. Data set for 2013, 2015, and 2018.

Comparing the manufacturer expected power output for the wind turbine and the actual generation capacity demonstrates a close relation between the theorical and real turbine capacity. To be able to have a flexibility in the combination of power to be installed four wind turbines have been selected with different generation capacities, hub height, and cut in and cut off wind velocities. Table 2 shows the basic data of the turbines.

Table 2. Technical data for Wind turbines [12].

Technical Data				
Specification	**HUMMER H13.2-20 kW**	**AERODAN 75/15**	**AN Bonus 150/30**	**ATB RIVA CALZONI 500.54**
Rated Power (kW)	20	75	150	500
Rotor Diameter (m)	13.2	17	23	54
Hub Height (m)	35	23	30/40	50
N°. Blades	3	3	3	3
Swept Area (m^2)	136.9	227.0	415.0	2290.0
Cut-in Wind Speed (m/s)	3.0	5.0	3.5	3.5
Cut-out Wind Speed (m/s)	25	25	25	25

Figure 10 depicts the fitted curve equation from the set of point from the manufacture power curve in the datasheet. The wind speed was measured at 10 m height and adjusted to the velocity as required by the hub of the wind turbine, from the set of turbines selected the wind speeds where adjusted to a hub height of 35 m and 50 m. This was calculated by applying the Hellman coefficient correction for the wind speed at hub height (Equation (3)) [6].

$$V_{h_2} = V_{10} \times \left(\frac{h_2}{10}\right)^{\alpha}, \tag{3}$$

where

V_{h_2} = velocity of the wind (m/s), at height h_2;
V_{10} = velocity of the wind (m/s), at height h_{10} = 10 m;
α = Hellmann coefficient, in this case 0.11.

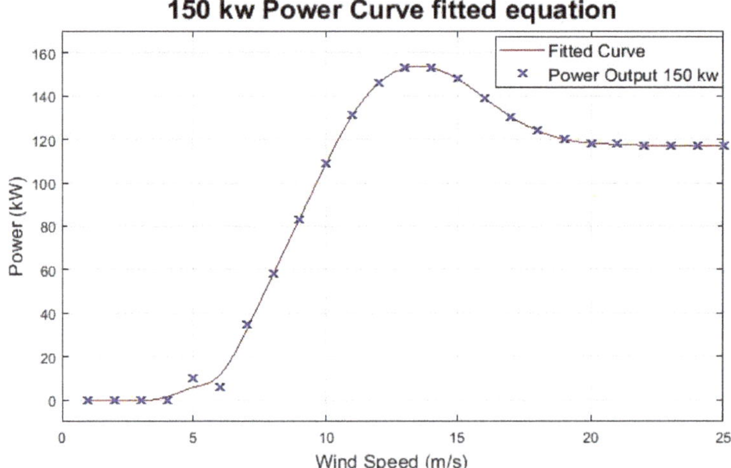

Figure 10. The 150 kW power curve fitted equation.

Finally, to calculate the power output of the wind turbine, Equation (3) was used and the new velocity at the hub height using the MATLAB function "Spline" to obtain all the individual values in 10 min intervals of recorded wind speed (Equation (4)).

$$P_{Wind} = \text{spline}\left(V_{turb}, Y150, V_{h_2}\right), \quad (4)$$

where

V_{turb} = known wind velocity from turbine power curve (m/s);
Y150 = known power output for the standard wind velocity (V_{turb}) (kW);
V_{h_2} = Sark calculated wind velocity at the hub height required.

Figure 11 shows the output for the 150 kW wind turbine with the result of Equation (4). As expected, in the summer days of the year, there is a lower wind speed close to the cut-in speed as can be observed within the red square (1) in Figure 11. As a final representation of the power output of the turbine, a representative two-day time span is plotted with the 10 min intervals data in the next graph for a typical day of summer and winter.

For the selected 150 kWp wind turbine system size, the annual generation is 483.5 mWh/year.

$$CF = \frac{\text{Wind annual energy generated}\left(\frac{\text{mWh}}{\text{year}}\right)}{\text{Rated peak power (kW)} \times 24\ \text{h} \times\ 365\ \text{day}} = \frac{483.5\ \frac{\text{mWh}}{\text{year}}}{1314.0\ \frac{\text{mWh}}{\text{year}}} = 36.79 \quad (5)$$

From this value, the capacity factor (CF) for the installation in Sark Island was calculated, resulting in a 36.79% capacity factor. The UK standard CF is 31% [13]. The higher value for this island is expected due to the location of the turbine where a CF closer to an offshore wind turbine is characteristic.

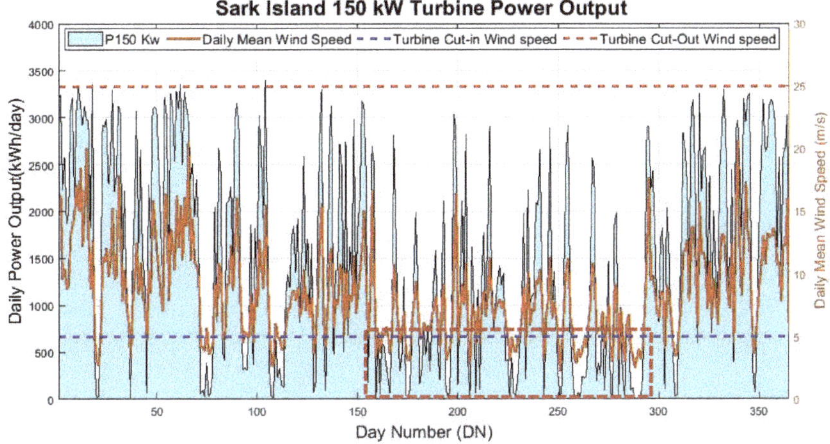

Figure 11. Sark Island 150 kW turbine power output.

4.3. Solar PV Generation

The data parameter needed to calculate the PV energy generation for a specific location have been acquired from the NASA [9] website utilizing the Sark Island latitude and longitude (49.432–2.36) [14]. Table 3 shows the data for a standard day in each month for the maximum and minimum temperature, the daily diffused (D) and global horizontal (G) irradiation.

Figure 12 shows Sark Island irradiance data from values of daily diffused (D) and horizontal (G) irradiation from Table 3 using the EXCEL program to split the data into hourly I_G and I_D.

Table 3. NASA annual temperature and irradiance values of Sark Island.

Parameter	Jan.	Feb.	Mar.	Apr.	May	Jun.	Jul.	Aug.	Sep.	Oct.	Nov.	Dec.
T_{max}	8.91	8.52	9.57	10.93	13.35	15.65	17.68	18.49	17.51	15.44	12.26	10.04
T_{min}	6.84	6.33	7.19	8.39	10.74	13.13	15.18	16.08	15.17	13.22	10.22	7.97
I_D	0.61	0.96	1.58	2.09	2.48	2.65	2.55	2.21	1.76	1.12	0.73	0.5
I_G	0.94	1.70	2.90	4.61	5.94	6.27	6.10	5.22	3.84	2.14	1.18	0.74

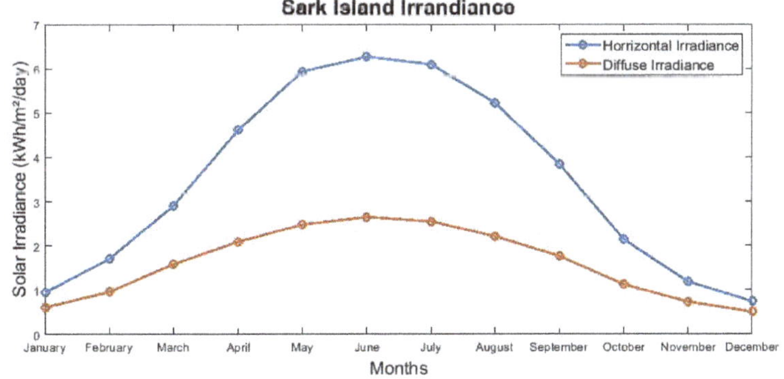

Figure 12. Sark Island irradiance [9].

To obtain the best possible approximation of the solar panel energy production accurate hourly data are required. In this case the hourly temperature equation (Equation (6)) was utilized in terms of a Z-parameter from an ASHRAE estimated hourly computer model [15,16].

$$T_h = (Z \times (T_{max} - T_{min})) + T_{min}, \tag{6}$$

where

T_h = temperature in any hour;
T_{max} = daily maximum temperature;
T_{min} = daily maximum temperature;
Z = ratio of hourly temperature variation.

The next step is to calculate the Slope Global Irradiance for each month on Sark Island using the values of diffused (D) and global (G) irradiation [9] shown in Table 3. In addition, a Tilt inclination of 25° was selected with a south orientation of 180° to optimize the solar energy absorption. In order to include the weather variation, the daily mean of three years' worth of data has been obtained and used to calculate a correction factor for each day. The corrected slope irradiance is shown in Figure 13.

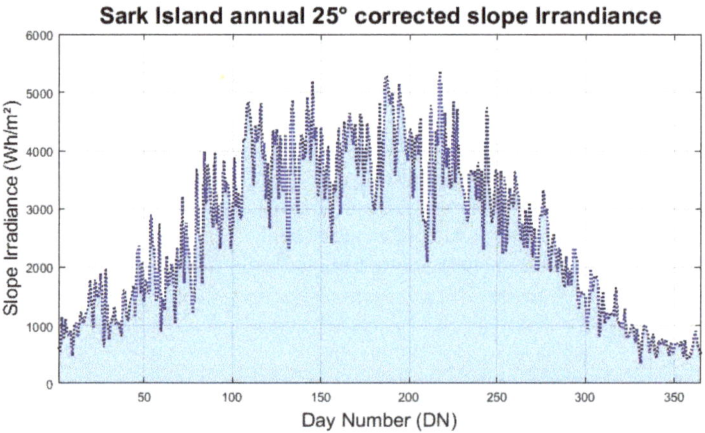

Figure 13. Sark Island corrected annual 25° slope irradiance.

Applying all the variables previously calculate for each hour of the day and taking into consideration the variation on efficiency due to temperature and sizing of the system. The monthly power output for a single PV panel had to be calculated (Equation (7)). The daily sum of the generation can be observed in Figure 14.

$$\text{Power out } (P_{out}) = \text{Cell Area } \times \text{ Global slope radiation} \times \text{ adjusted cell efficiency} \tag{7}$$

The adjusted cell efficiency, as defined by the Equation (8), which varies depending on the cell temperature using the temperature coefficient (α_p).

$$\eta_{cell} = \eta_{stc}\left[1 + \alpha_p(T_C - T_{c,stc})\right], \tag{8}$$

where

η_{cell} = solar cell temperature efficiency;
η_{stc} = efficiency under Standard Test Conditions and is 18.5%;
α_p = temperature coefficient;
T_C = cell temperature;
$T_{c,stc}$ = cell temperature under standard test conditions.

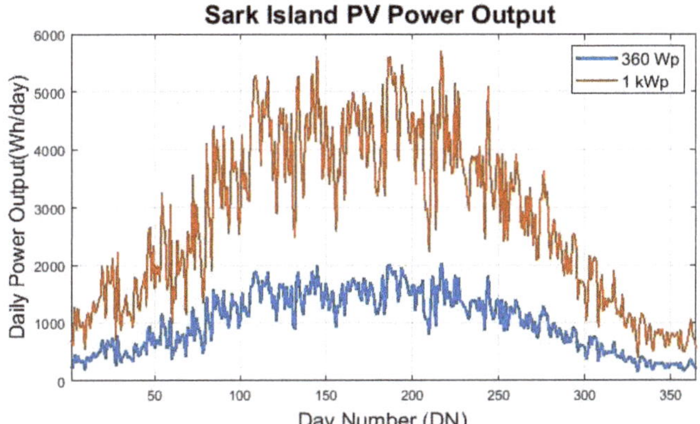

Figure 14. Sark Island PV power output.

The results for every hour in a month reveal a variation from the datasheet efficiency of 18.5%. The cell efficiency varies from 19.20% to 21.63%. The solar data has been divided from one-hour intervals to 10 min to standardize all the data sets. Figure 15 presents two days of summer (summer solstice—20 June to 21 June).

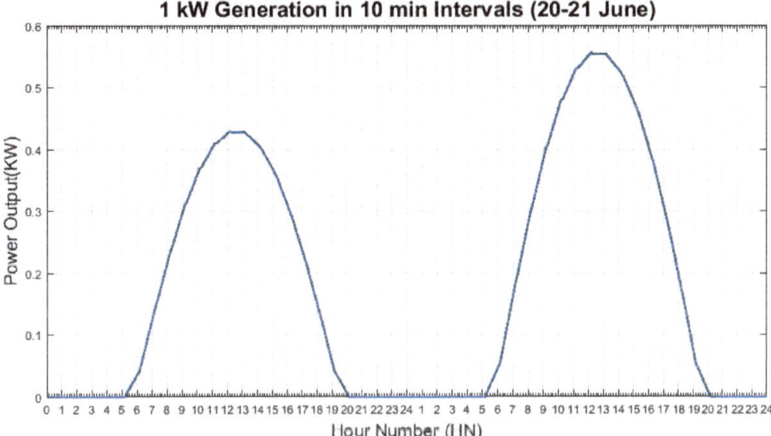

Figure 15. 1 kW Generation in 10 min intervals (20–21 June).

For the selected 1 kWp PV system size the annual generation is 1025.0 kWh/year.

$$\mathrm{CF} = \frac{\text{Solar annual energy generated}\left(\frac{\mathrm{mWh}}{\mathrm{year}}\right)}{\text{Rated peak power (kW)} \times 24\ \mathrm{h} \times\ 365\ \mathrm{day}} = \frac{1.025\frac{\mathrm{mWh}}{\mathrm{year}}}{8.760\ \frac{\mathrm{mWh}}{\mathrm{year}}} = 11.70\ \% \tag{9}$$

From this value, the capacity factor (CF) for the installation in Sark Island was calculated, resulting in an 11.70% capacity factor. The UK standard CF is 10.7% [17]. The higher value for the island is expected due to lower altitude, translating into more irradiance per meter square.

5. Cases of Study (Section III of the Methodology)

To find the best solution for the Sark Island actual energy requirements, a set of three different main cases are going to be compared for performance and integration of the renewables energy, the estimated size of the system, installation cost, and the CO_2 emissions.

5.1. Case 1: 100% Renewable Penetration

Figure 16 shows a simplified connection diagram of Sark Island, where the blue dotted square shows the existing diesel generators. The criteria to size this system was to be able to supply 100% of the energy required by the island just with a mix of renewable energy. This article only contemplated wind and solar as energy sources. The first step was to change the penetration of wind by using the 4 turbines output and add the required energy missing with the output of the 1 kWp PV system to size the solar system as well to supply 100% of the annual energy of the load (1600 MWh/year) (Table 4).

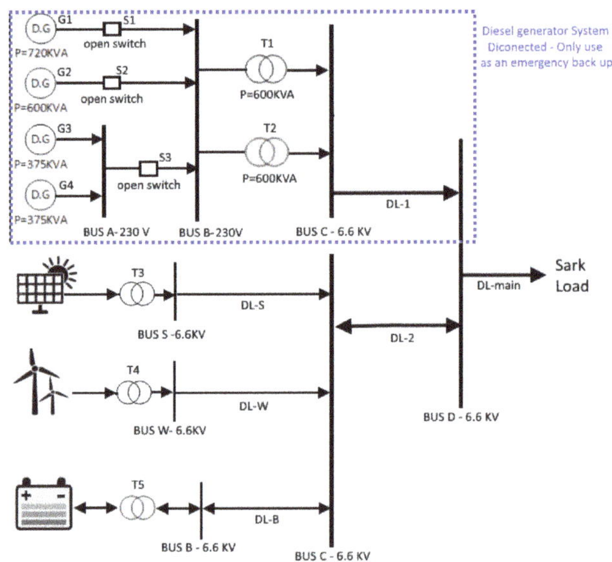

Figure 16. Case 1. Simplified proposed connection diagram of Sark Island.

Table 4. Case 1: System size and production results.

Case 1: Energy Balanced							
Scenario N°.	WTC (kWp)	WTAEG (mWh/year)	WEI (%)	ESPVC (kWp)	SPVAEG (mWh/year)	PVEI (%)	SIAEC (MWh/Year)
1	40	211.42	13.2	1579.6	1388.58	86.8	1600.0
2	115	330.15	20.6	1444.5	1269.85	79.4	1600.0
3	150	555.15	34.7	1188.6	1044.85	65.3	1600.0
4	225	779.59	48.7	933.3	820.41	51.3	1600.0
5	245	885.30	55.3	813.0	714.70	44.7	1600.0
6	300	1110.29	66.4	557.1	489.71	30.6	1600.0
7	320	1216.00	76.0	436.8	384.00	24.0	1600.0
8	435	1514.99	94.6	97.5	85.71	5.4	1600.0
9	450	1665.44	104.1	0.0	0.00	0.0	1600.0
10	500	2425.84	151.6	0.0	0.00	0.0	1600.0

WTC = Wind turbine capacity; WTAEG = Wind turbine annual energy generation; WEI = Wind energy integration; ESPVC = Estimated solar PV capacity; SPVAEG = Solar PV annual energy generation; PVEI = PV energy integration; SIAEC = Sark Island annual energy consumption.

With the size of the 10 different scenarios, the available energy at any given hour has been calculated at hourly intervals by deducting the instantaneous load from the available renewable energy, as shown in the next equation:

$$\text{Energy Balance (kWh)} = \left[\left(E_{wind} + E_{pv}\right) - E_{load}\right], \tag{10}$$

where

E_{wind}: Wind energy generation (kW);
E_{pv}: Solar PV energy generation (kW);
E_{load}: Sark Island consumption (kW).

After, the energy balances with one-hour intervals have been added to calculate the available energy to be stored in the battery (Equation (11)).

$$\text{Battery Energy(kWh)} = \left[E_{bal1} + E_{bal2}\right], \tag{11}$$

where

E_{bal1}: Initial Energy balance (kW);
E_{bal2}: next Energy balance (kW).

Then, the net energy for the battery has been calculated to have an annual net energy charge and discharge rate for the system. Figures 17 and 18 show the results for each of the renewable energy mixes.

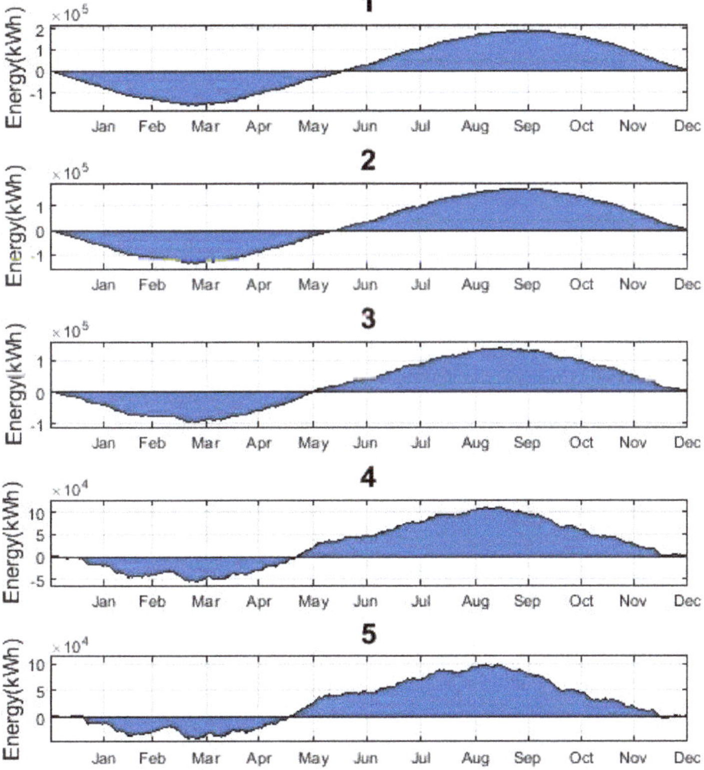

Figure 17. Part 1: Energy balance per scenarios 1–5.

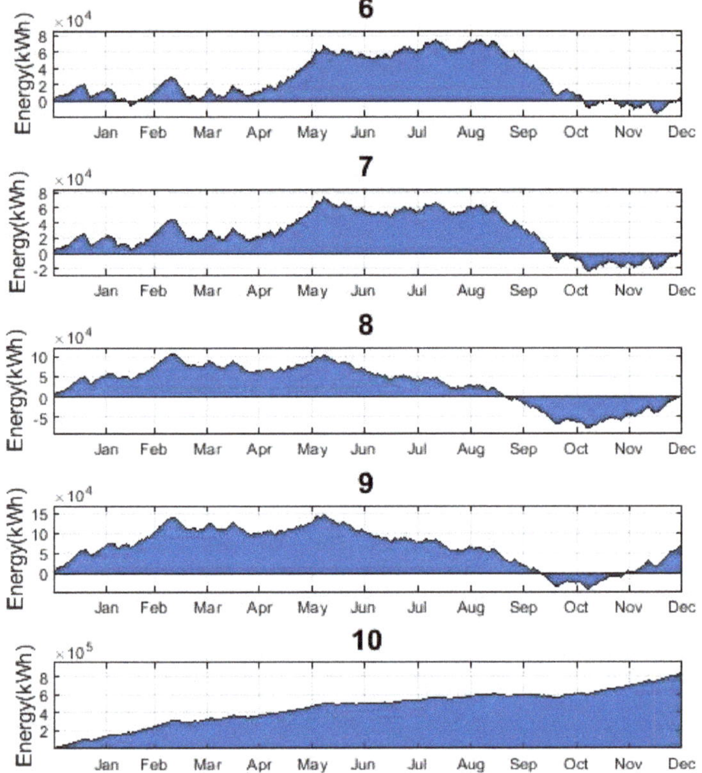

Figure 18. Part 2: Energy balance per scenarios 6–10.

The results of these graphs showcase the charge and discharge rate expected for the characteristic load of the island considering the specific renewable energy mix. The first five graphs have predominant solar generation, which create a characteristic behavior of discharge rate on the winter days sections 1 and 3 and charge rate on the months with higher irradiance Section 2 (Figure 19).

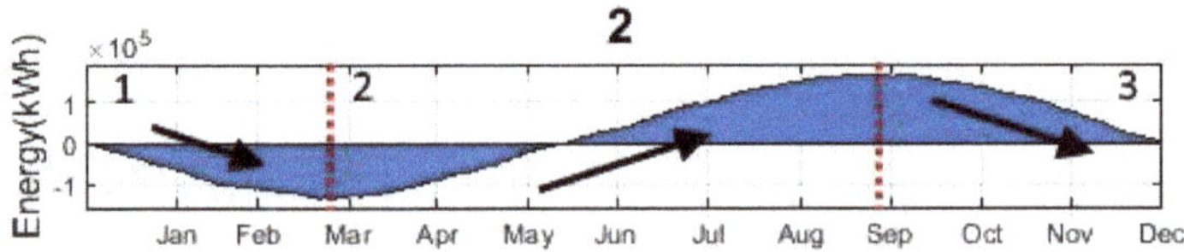

Figure 19. Scenario 2: Energy balance.

On the other hand, a higher percentage of wind generation is shown to achieve a less demanding energy system by reducing the time of constant discharge of the overall system. In scenario 6, a combination of 69.4% of wind and 30.4% of solar generation results in smaller estimated battery storage to deliver the energy during discharge periods. This assumption is based on the overall net energy use in sections 1–3–5 and the much smaller discharge rate state in Section 4, as shown in Figure 20.

In scenarios 7–9, the largest wind integration changes the time of discharge rate in summer days, as shown in Section 3 of the Figure 21. The charge periods in sections 1 and 4 occur at higher constant wind speeds.

Finally, in scenario 10, a constant charge rate occurs due to the oversizing of the system with a 150% generation integration compared to the load. This scenario will have the smallest possible battery storage needed at the cost of having over generation and energy loss (Figure 22).

After evaluating the system performance, the battery storage was designed on the worst-case scenario (Figure 23). Each scenario has been compared to select the most ideal system for the island utilizing the system configuration of Case 1.

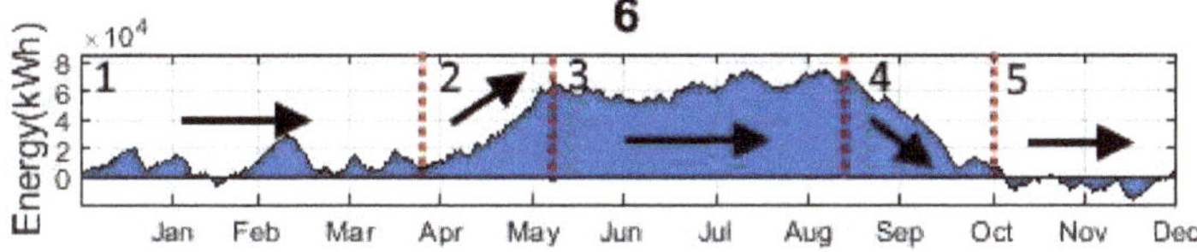

Figure 20. Scenario 6: Energy balance.

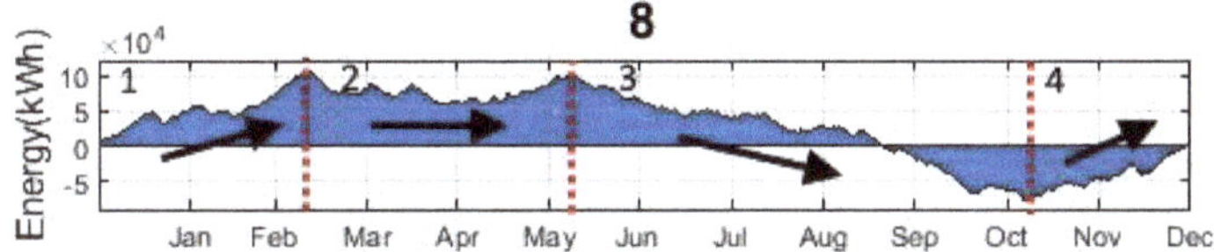

Figure 21. Scenario 8: Energy balance.

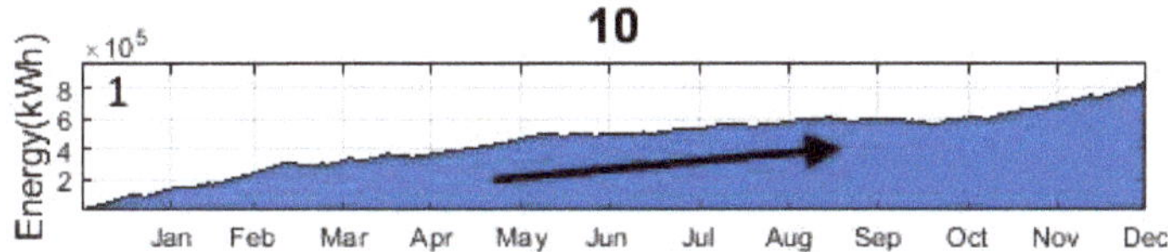

Figure 22. Scenario 10: Energy balance.

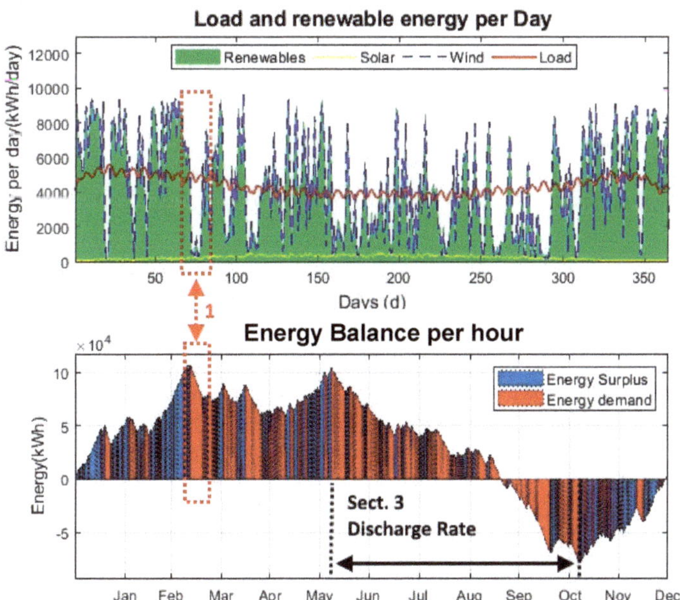

Figure 23. Scenario 8: Annual load and renewable generation.

5.2. Case II: Renewable Energy Generation and Battery—GenSet System

The second case study considers a mix or renewable energy generation and a battery storage system (BSS) along with the existing diesel generation, which is necessary to reduce the size of the battery bank in order to decrease the installation cost. The use of the small generator currently available on the island could supply energy when the renewable sources are insufficient to supply the load and recharge the batteries (Figure 24).

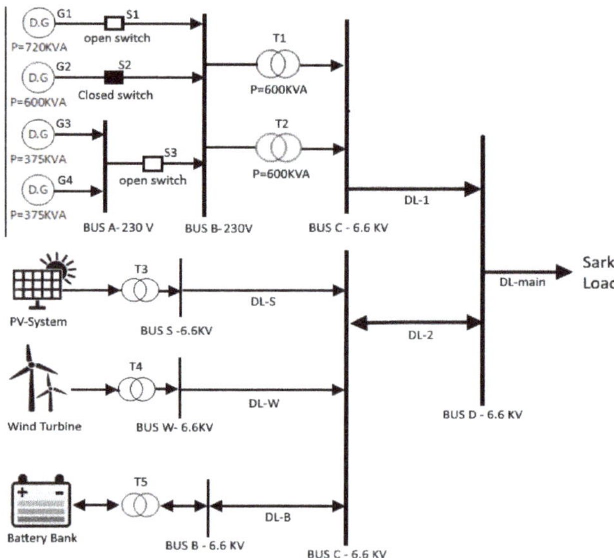

Figure 24. Case 2. Simplified connection diagram of Sark Island.

A minimum autonomy of one day (24 hours) has been assumed for the battery as an initial design point along with a minimum SOC of 40% and a daily mean consumption for a day (E_{req2}) of 4.38 MWh/day whilst, the same overall battery system efficiency was assumed as

$$\text{New Battery Size (MWh)} = \frac{E_{req2} \times (1 + SOC)}{e_{syst}}. \quad (12)$$

The results for Equation (12) are a new battery bank of 7.22 MWh, with a reduction in size of 92% compared to the average size of the case 1 results. From the methodology section, a peak load consumption of 295 kW was obtained and used to size the inverter considering an overrating of 1.3 times thus sizing the inverter at 383.5 kW. With the selected inverter and the battery size of the system, the energy storage system of Autarsys has been selected. A system efficiency for charging and discharging the battery storage of 85% [18] has been assumed.

As shown in Figure 25 the peak hours of consumption are between 8:00 and 23:00. The code has been based on the same structure of the energy balance calculation for case 1 while adding the condition of charging the battery at the minimum state of charge. The latter has been limited by the hours the generator starting when the SOC is less than 50% and staying on until the battery is fully recharge.

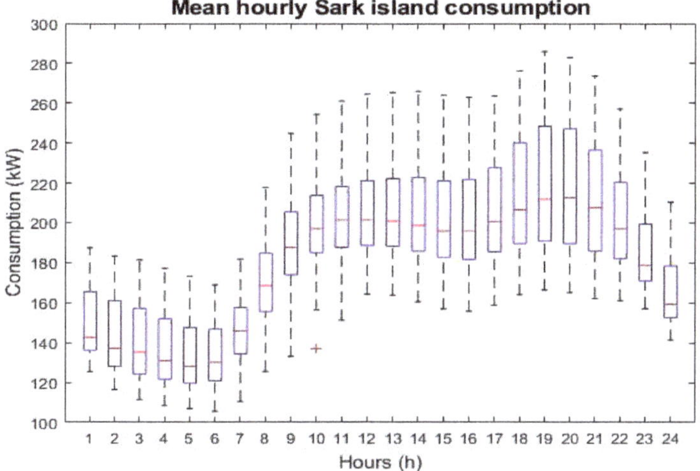

Figure 25. Mean hourly Sark Island consumption.

The best scenarios from the previous case (scenarios 6 and 10) have been re-evaluated in case 2 with the same battery storage capacity, while three more additional battery sizing scenarios have been evaluated to compare the difference in cost and incidence of diesel generation.

The new scenarios are divided as follow (Table 5):

Table 5. Case 2. Scenario analysis.

Renewable Energy Systems		Scenario N°.	Battery Size
Case 1. Scenario 6	**557 kWp PV and 300 kW Wind**	1	0.90
		2	1.80
		3	3.61
		4	7.22
Case 1. Scenario 10	**500 kW Wind**	5	0.90
		6	1.80
		7	3.61
		8	7.22

Only one scenario is hereby presented (Case 2, Scenario 6: 500 kW Wind + 1.80 MWh battery storage) in order to avoid repetition and plotting similar figures with small variance.

Figure 26 shows the overall renewable energy surplus (waste) in comparison with the total amount of energy required to recharge the battery by the generator. A distinctive behavior can be observed. Due to reliance on wind generation an energy surplus is predicted in the coldest months, while between May to October, the presence of the diesel generator is higher due to lower wind speeds.

The more representative month of October, chosen as a transitional month, has been plotted in Figure 27 to provide further insight. In this month, the use of the generator to recharge the battery is required at least twice per day for the worst day of renewable energy generation. This is important as the cycling of the battery will affect the expected battery life of the system. Furthermore, the necessity of the diesel generator is highlighted for recharging during those intervals due to insufficient renewable energy. Finally, the annual energy balance for each scenario is compared in Table 4 and presented as a comparative plot in Figure 28.

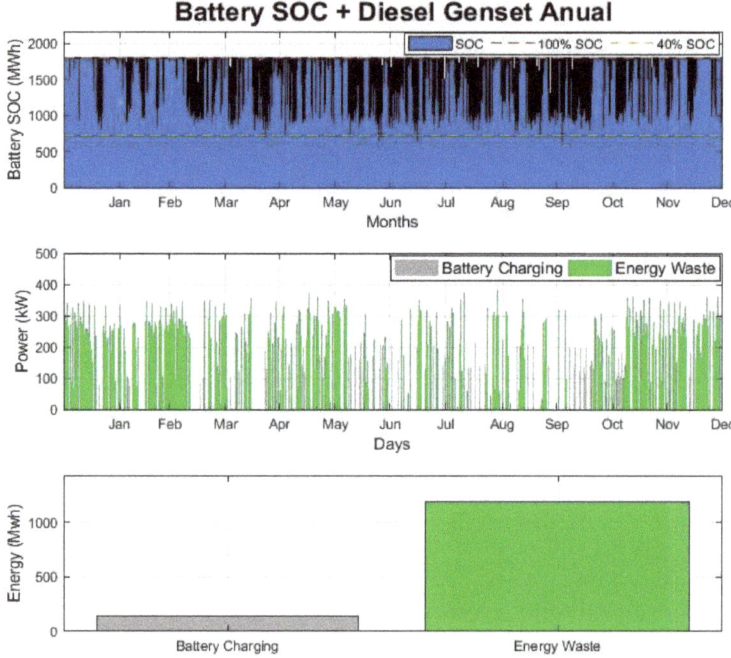

Figure 26. Scenario 6: 500 kW wind turbine and 1.80 MWh battery storage annual results.

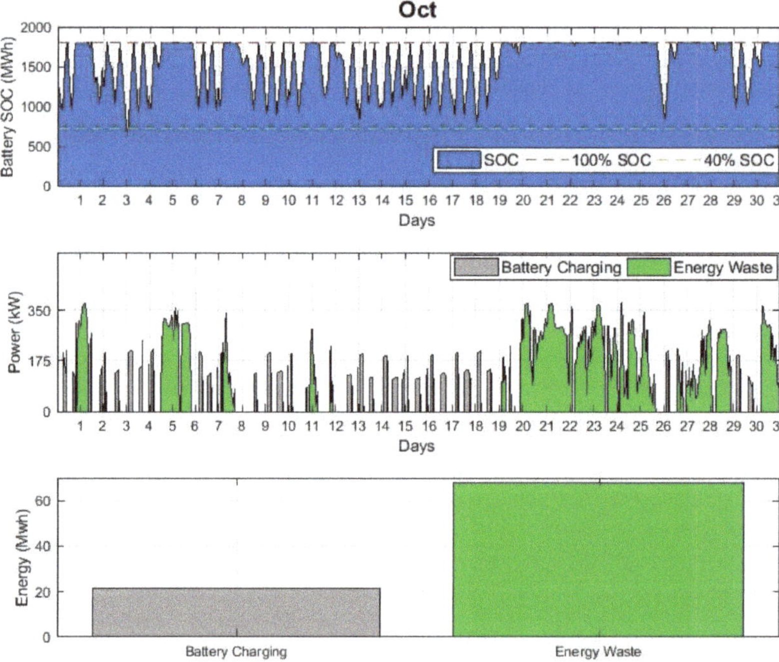

Figure 27. Scenario 6: 500 kW wind turbine and 1.80 MWh battery storage (October).

The total renewable energy generation for each case can be observed along with the total energy waste (surplus) and the required diesel energy generation, thus summarizing the results. Comparing the two renewable generation systems, an average of 18% of the energy use in the system comes from the generator. The notable difference in this analysis is the average energy waste between the two energy systems. Case 1, scenario 6 has a 27% of energy waste in the year, versus the 73% was from the case 2. The amount of cycles is dictated by the size of the battery bank and not by the renewable energy generation profile. From this, the smaller size 0.9 MWh of storage will cycle four times a day, resulting in a short expected battery life of approximately seven years. This translates into the need of three battery bank replacements and re-investment in the 20 years analysis. For the next battery size of 1.80, the expected battery life would be approximately 14 years, so it will need two installations. Finally, the last two battery sizes have 1 and 0.5 cycles per day, which results in an expected battery life of more than 20 years (Table 6).

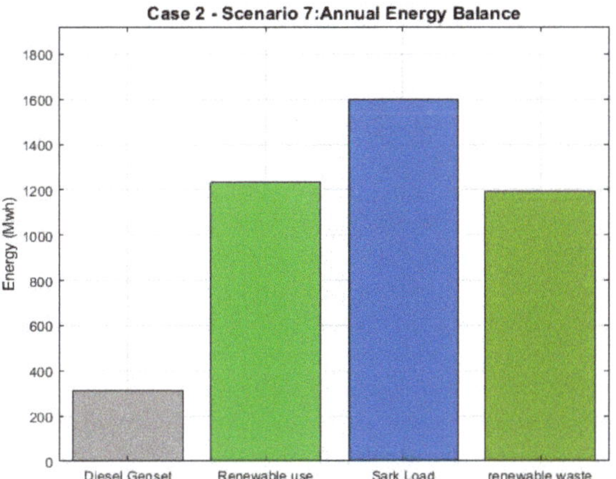

Figure 28. Case 2, Scenario 7: Annual energy balance.

Table 6. Case 2: Renewable generation, diesel output, and battery performance per storage size.

RS	Scenario N°.	BS (MWh)	SIC (MWh/year)	RG (MWh/year)	EW (MWh/year)	GD (MWh/year)	CPY	GAU (Hours)	WDAC
Case 1. Scenario 6.	1	0.90	1600.0	1681.18	518.89	285.7	354	866	4.0
	2	1.80	1600.0	1681.21	489.85	332.2	290	1,007	2.0
	3	3.61	1600.0	1681.29	414.30	291.3	266	883	1.0
	4	7.22	1600.0	1681.37	331.43	229.6	184	696	0.5
Case 1. Scenario 10	5	0.90	1600.0	2424.60	1245.90	286.1	287	867	4.0
	6	1.80	1600.0	2424.65	1194.43	312.1	245	946	2.0
	7	3.61	1600.0	2424.69	1152.20	307.8	175	933	1.0
	8	7.22	1600.0	2424.76	1083.22	255.7	156	775	0.5

RS = Renewable system; BS = Battery size; SIC = Sark Island consumption; RG = Renewable generation; EW = Energy waste; GD = Genset-Diesel; CPY = Cycles per year; GAU = Generator annual use; WDAC = Worst day amount of cycles.

5.3. Case III: 40% Renewable Penetration

For this case, the connection of the system considers one of the smaller generators of 375 KVA working in parallel with the renewable generation system without requiring the use of a battery bank (Figure 29).

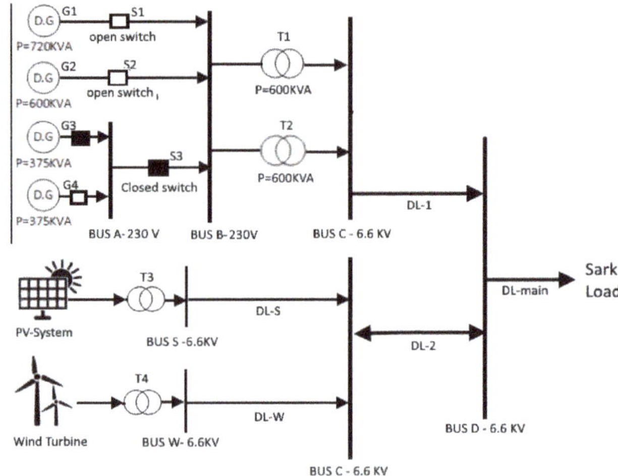

Figure 29. Case 3. Simplified connection diagram of Sark Island.

In the final case study, the renewable energy system was sized to supply an average 40% of the load with a wind-only renewable generation supply. This assumption is based on the requirement of a minimum constant output of the diesel generator (in this case, 20% of the nominal load) to work under a high efficiency of 92%. Additionally, it can provide ancillary services as a grid stabilization as well as supplying the load when a shortage of renewable energy is present. Furthermore, from the results of case 1 and the load analysis for case 2, a combination of wind and solar generation systems is more suitable for the weather conditions of the island, the estimated consumption, and the energy profile. This is because of the higher demand during cold months, where the wind generation is more stable and solar could aid in the reduction of daytime energy demand peaks. It is important to note that this selection has been purely based on the estimated load profile assumed for the island. Considering the island conditions and the relative price of renewable generation, a trend is revealed toward lower prices for higher energy integration in the system. Thus, a 75% wind integration and 25% solar integration is presented for this case study.

The wind turbine has been sized at 150 kW, generating on average 555.15 MWh/year, and the solar PV at 150 kWp of installed capacity generating 153.84 MWh/year, of which 6% is expected to be lost as over generation (surplus), resulting in a net renewable energy use of 620 MWh/year, this being approximately 39 % of the total energy supply (Figure 30).

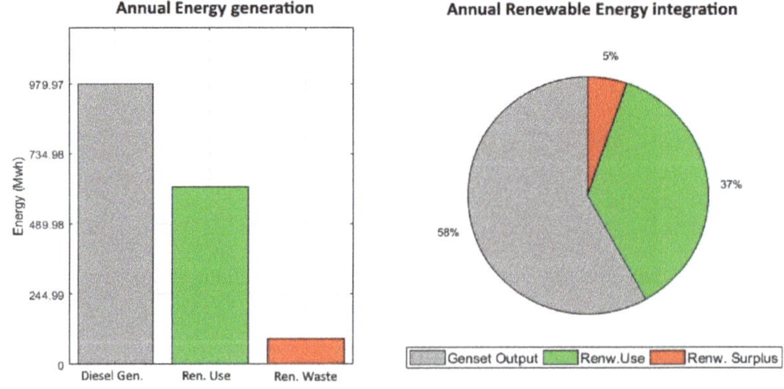

Figure 30. Annual energy generation and integration.

To assess the generation behavior for the best-case and the worst-case scenarios, a monthly energy balance percentage has been plotted representing the renewable integration for each system and the energy waste (Figures 31 and 32).

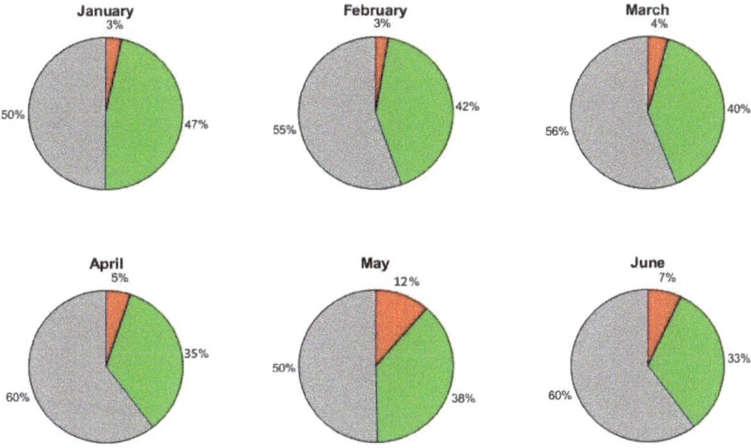

Figure 31. Case 3: First half of the year energy balance.

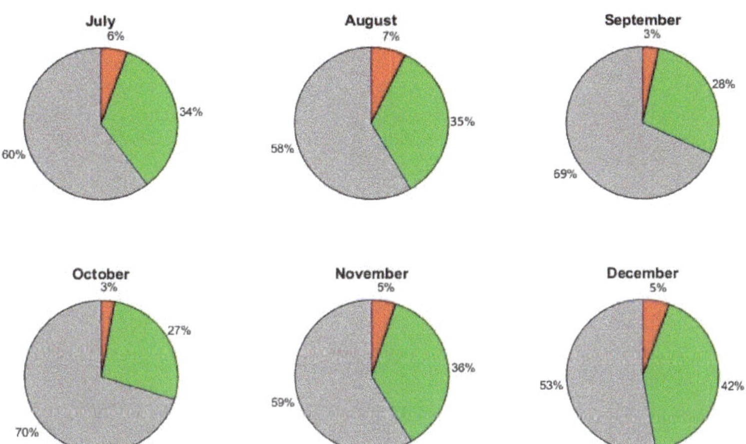

Figure 32. Case 3: Second half of the year energy balance.

The results show that the worst month for renewable energy generation would be October (with the lowest energy losses). The month with highest generation and energy integration is the month of January. It is important to notice that, even though this month requires more diesel generation, the energy surplus is lower than the month of December due to the increase of the load in the month of January.

For the month of January, the wind speed is fairly stable, with just a couple of day with speeds lower than the cut-in speed of the turbine, thus having a renewable energy output of effectively 0 kW. This has been taken into account, with the diesel generator previously working only at 20 % of its nominal capacity. During the remaining days, the wind generation is enough to supply the load and have an average 3% energy surplus. Another important aspect for this month is the supply of the load by renewable energy generation with an integration of 47% of the consumption. The energy currently wasted in this design could be reduced if the load increases in the following years. Also, the biggest energy surplus occurs in the night hours; this is related to the energy profile estimated for this island, where the biggest consumption peak occurs in the day (Figure 33).

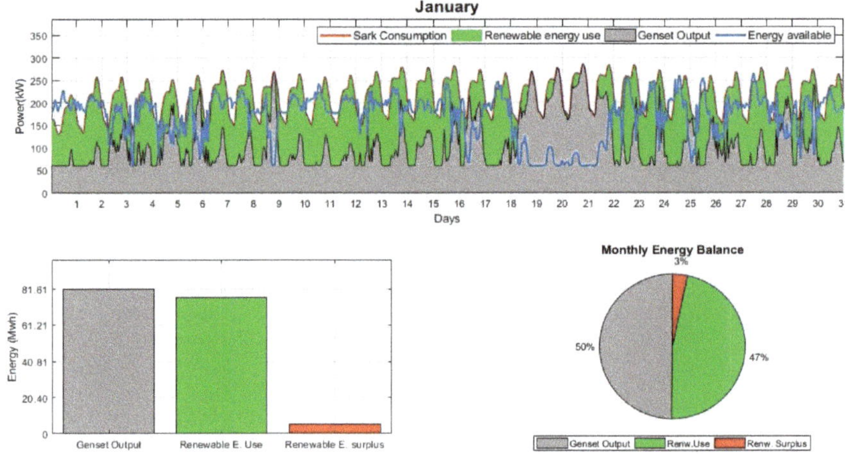

Figure 33. Case 3: January energy balance.

The month of October has a bigger diesel generation ratio with 78% cover of the load and an almost constant low generation with more than five consecutive days without any generation. A potential solution for this month would be the integration of solar generation, but as explained in previous cases, the sizing of the PV system would have to be oversized to have a significant improvement (Figure 34).

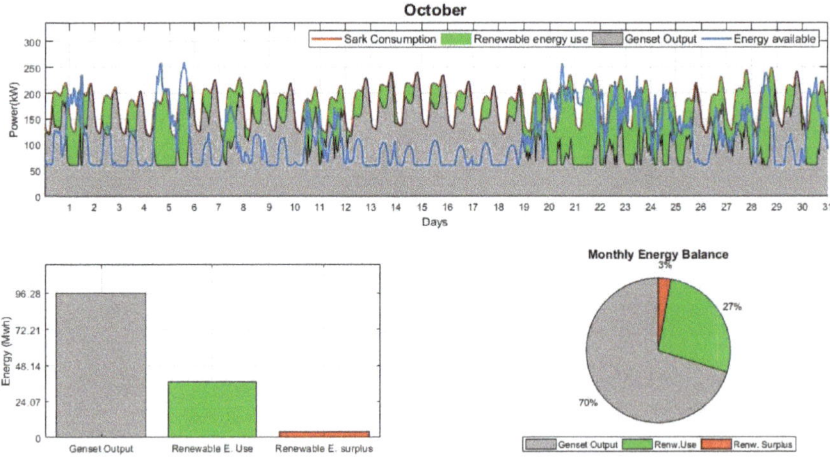

Figure 34. Case 3: September energy balance.

6. Economical and CO_2 Emissions (Section IV of the Methodology)

6.1. Case I: 100% Renewable Penetration

The installation cost for each scenario can be calculated and compared as well as the CO_2 emissions (Figure 35).

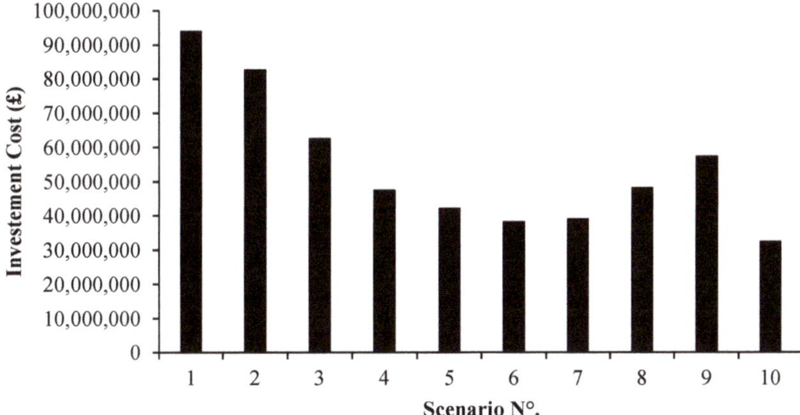

Figure 35. Case 1: Installation cost and O & M (Operations and Maintenance).

In case 1, the most economical system able to supply 100% renewable energy generation and distribution is scenario 10—installing a 500 kW wind turbine with no solar integration. It is important to note that, in case 6, the second lowest installation cost is the energy balance of approximately 70% wind and 30% solar generation.

For this case study, the results, shown in Figure 36 reveal that the CO_2 emissions are almost similar between all the scenarios, with an average of 58 gCO_2/ kWh and a maximum deviation of 12% for the output obtained in scenario 9. This is because of the massive size of storage required to supply the load in the summer months.

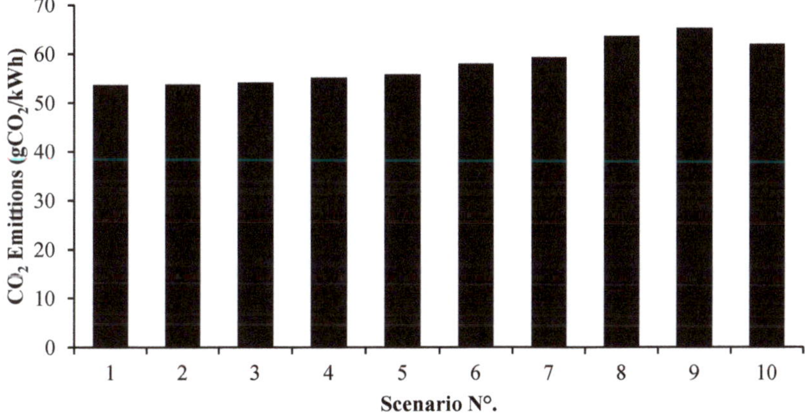

Figure 36. Case 1: CO_2 renewable energy mix emissions.

If we only take into account the cost of installation, the solution would be to install the oversized system in scenario 10, with a 500 kW wind turbine resulting in an annual over generation of 151%. This reduce the battery bank size to the minimum value for this condition due to the positive energy balance throughout the year, as shown in Figure 22.

The battery size to be able to deliver a 100% renewable system with only a combination of PV, wind and battery storage results in the requirement of a battery bank that is both oversized and features extreme rates of discharge during prolonged periods resulting in the battery installation cost being approximately 98% of the total investment. For this reason, in the second case study, the generation system in scenario 6 has been combined with the existing diesel generator on Sark Island to reduce in size the required battery storage. Scenario 6 has been chosen as an initial design point due to the lowest installation cost and the most stable energy balance with the lowest discharge rate period, as shown in Figure 20.

The scenario also allows for comparison of the benefits of a more distributed energy generation between summer and winter, reducing the necessity of over generation.

6.2. Case II: New Battery Sizing

The results in this case reveal that the difference among the installation cost from the smaller system is less than 17% but having the additional difficulty due to the need to change the battery bank every seven years, resulting in an unsuitable system (Figure 37). The installation difficulty due to the additional cost of transportation required has not been contemplated in this comparison. The ideal system solution from the results is a battery size of 3.61 MWh because of the expected storage battery life of 20 years.

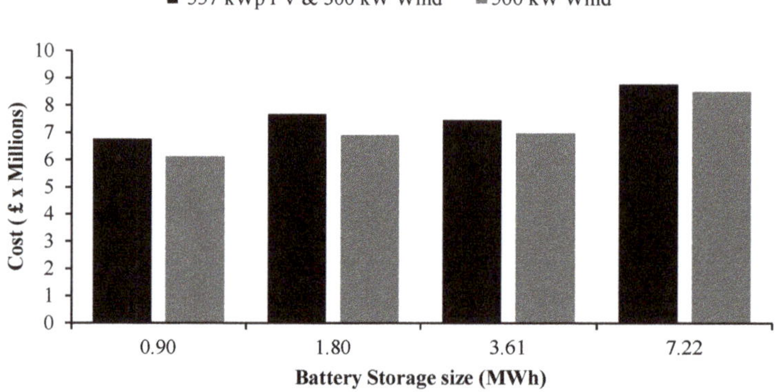

Figure 37. Case 2: Installation cost for the overall system.

For this case study, the results, shown in Figure 38, reveal that the CO_2 emissions are similar between all the scenarios, with an average of 140.1 gCO_2/ kWh and a maximum deviation of 22% for the output obtained in scenario 2. This is because of the additional use of the diesel generator compared to the other scenarios.

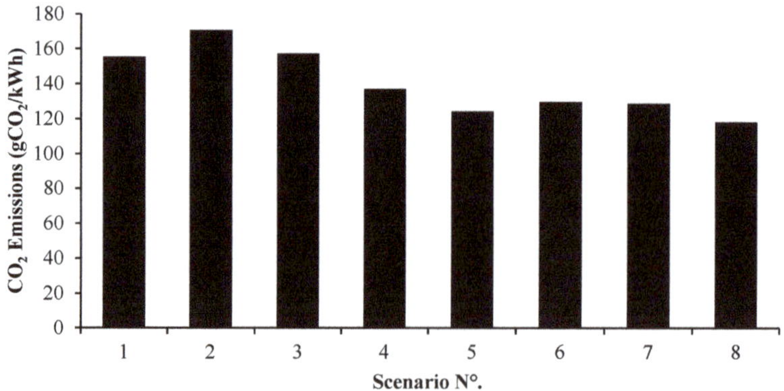

Figure 38. Case 2: CO_2 renewable energy mix emissions.

6.3. *Case III: 40% Renewable Penetration*

As a final case evaluation, the scenarios with the best performance for each individual case have been assessed in order to compare the different advantages and disadvantages for each solution.

When only evaluating the installation cost, case 1 makes the investment cost extremely high due to the size of the battery bank. Compared with the second case this is more than 10 times the cost. Additionally, the difference in cost for case 2 and case 3 is insignificant when compared to case 1. In order to be able to properly compare the scenarios' economic feasibility, the cost of O & M in a 20-year analysis was required. From this, it can be shown that the diesel operational cost is going to be around 14 million pounds for 20 years. With a simple analysis, case 1 can be discarded due to high installation cost, while it is assumed that all the diesel running cost will return just 50% of the initial investment. Finally, if the same simplified return on investment is applied to cases 2 and 3, a 100% return of investment is obtained at 5.8 years for case 2 and 2.58 years for case 3.

CO_2 emissions for the current energy generation is estimated from the average GHG emission of a diesel generator to around 458 gCO_2/kWh. For case 1 a reduction of to 86% can be expected compared to case 2, where the reduction will be of 71%. For the case 3, with the lowest investment cost, a high emission rating is expected due to the continuous use of diesel as the main energy source, reducing just 20% of the actual emissions (Figure 39).

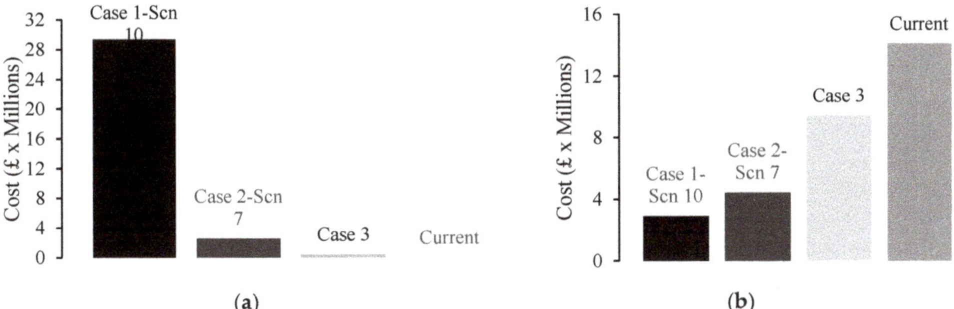

Figure 39. Case 3: Installation cost comparison (**a**); Case 3: 20-year O & M (**b**).

7. Discussion

7.1. *Case 1: 100% Renewable Penetration*

If only taking into account the cost of installation, the solution would be to install the oversized system in scenario 10, with a 500 kW wind turbine resulting in an annual over generation of 151%. This reduces the battery bank size to the minimum value due to the positive energy balance throughout the year, as shown in Figure 22.

The battery size to be able to deliver a 100% renewable system with only a combination of PV, wind, and battery storage resulting in the requirement of a battery bank that is oversized and with extreme rates of discharge during long periods of time, resulting in the battery installation cost being approximately 98% of the total investment. For this reason, in the second case, it will be combine with the generation system in scenarios 6 and 10 with the help of the diesel generator on Sark Island to reduce in size the required battery storage.

The scenario 6 has also been chosen for the second case study to compare the benefits of a more distributed energy generation between summer and winter and reduce the occurrence of over generation alike in scenario 10. One reason to choose scenario 6 is that is the second with the lowest installation cost and the most stable energy balance with the lowest discharge rate period as shown in Figure 20.

7.2. Case 2: Renewable Energy Generation and Battery—GenSet System

From the results in Table 6, the most optimal renewable energy system generation can be achieved in scenarios with a 500 kW wind turbine due to the similar use of the diesel generator, but having an advantage of more than 73% of energy surplus that can be utilized in the future if the load increases. In addition, scenario 7 has an ideal ratio between cost of installation and expected battery life. To conclude, with CO_2 emissions being nearly double those of case 1 results, the emissions are still 70% lower than the actual CO_2 emissions level. Scenario 7 has 120 gCO_2/kWh, which is 7% lower than the average for this system.

7.3. Case 3: 40% Renewable Penetration

CO_2 emissions for the current energy generation is estimated from the average GHG emission of a diesel generator to around 458 gCO_2/kWh. For case 1, a reduction of to 86% can be expected. For the Case 2, the reduction will be of 71%. Case 3 has the lowest investment cost and still has a high emission rated due to the continues use of diesel as the main energy source, reducing just 20% of the actual emissions (Figure 40).

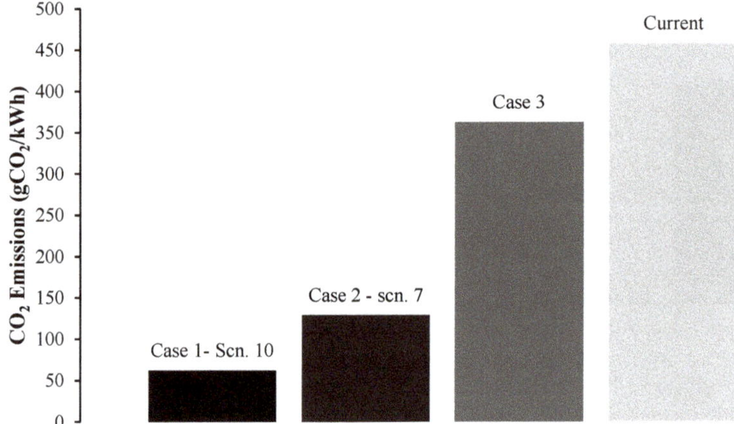

Figure 40. CO_2 emissions.

The current LCOE from the island is 0.66 £/kWh comparing to this which scenarios have the best expected economic performance based on the installation cost, O & M, and the total energy generation (Figure 41). The degradation on energy production for solar PV and wind was taken into account. This gave the following results: case 3 was 30% higher than the actual energy cost, meaning that the electricity will need to be sold at a higher price than the current price in order to generate earnings.

Furthermore, the same will occur with the electricity price for the case 1. On the other hand, case 2 shows a reduction of energy production cost of 33% to a minimum value of 0.42 p/kWh. This will enable a reduction of the selling price of the electricity.

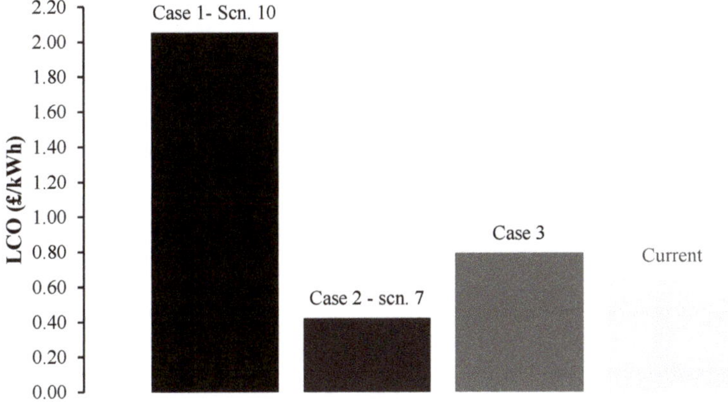

Figure 41. Lowest levelized cost of energy (LCOE).

8. Conclusions

The main objective for this project was to design an energy system after analyzing different alternatives of renewable energy integration to reduce the current electricity cost to 52 p/kWh from August 2018 to 2019 and then reduce it to a minimum of 49 p/kWh from 2019 to 2020. One of the main requirements at the start of this project was to be able to obtain an annual load profile from the island itself, but after conducting research, we determined that this was not possible due to lack of available data. The annual load has been estimated as an hourly consumption profile for Sark Island, which directly influenced the results of this study by adding currently unknown uncertainty to the calculations. Because of this, the results can vary if real data is obtained and the assumed consumption profile is different. As an example, the load profile selected resembles the behavior of the UK mainland, with peak consumption in winter due to the similar average temperatures in the year, but it was also valid to assume a higher demand in the summer months due to an increase in tourists, just as the NAREC did in their report [19]. This report presents the case for electricity generation for the island, where a mix of solar and wind is used for the worst-case scenario, focusing on the estimated monthly energy consumed. The system sizing was undertaken by oversizing to reduce the risk of not enough energy being available. The method implemented in this research takes this into account and combines the current diesel generators to be able to reduce the uncertainty. By reducing the uncertainty, it has been possible to reduce the size of the solar system by 80% and both the wind turbine capacity and battery size by 50%.

After the data analysis, assumptions and the different sets of results for the installation cost, carbon emissions, and levelized cost of energy, the findings of this study can be summarized as follows.

- Case 1: A purely wind energy generation (500 kW) and an oversized battery bank can obtain the minimum greenhouse gases emissions out of all the cases, with a reduction of 90%, but this requires a high investment cost due to the still elevated price of the battery storage. In this scenario, a battery size of approximately 54 MWH is required. Also, in the 20-year LCOE study, this case results in a 2.1 £/kWh energy cost. This is not suitable as a solution for this island, as the cost of energy would be three times higher than the current price.
- Case 2: In this case, the same wind turbine has been used rated at 500 kW, but with the addition of using the small generator of 375 KVA to charge the battery when renewables are not enough. This enables a drastic reduction of the battery bank to a 3.61 MWh of storage or 12 hours of autonomy of constant discharge. Also, the GHG emissions are 70% lower than the actual CO_2 emissions level, with a value of 120 gCO_2/kWh. Finally, this case shows the most economical feasible LCOE, with a reduction on the energy production cost of 33% to a minimum value of 0.42 p/kWh.

- Case 3: The final case used a mix of solar and wind to have a more stable energy output during the day, when the peak on the load occurs with a 150 kW wind turbine and 150 kWp solar-PV system, producing 39% of the total annual load—75% comes from wind and 25% from the solar arrangement. This set up ends with an energy surplus of 6%. For this case in particular, the battery storage was eliminated in order to reduce the initial installation cost, and instead, the grid stabilization relies on the diesel generator of 600 KVA working constantly at a minimum base generation of 20% is nominal capacity and supplying the load when the renewable are not enough. This system has a minimum installation cost but elevated O&M and GHG emissions, with a reduction of only 20% on the emissions. Finally, this set up will have a higher cost of energy than the actual by 33%.

After summarizing all the results of this research and as previously explained, the results of this report depend vastly on the assumptions realized at the research and calculation stage. However, the case 2 renewable energy generation system is the most suitable in terms of the reduction of CO_2 emissions and expected earnings from a lower LCOE. This study can be re-evaluated if actual data from the island is available, following the same methodology, in order to find the best solution for the island's current energy generation problem.

As Supplementary Material, a new load profile has been created where the load reaches a peak during the summer, in contrast to the previous assumption where the peak took place in the winter. This is based on the assumption that the large influx of visitors on the island will have an equivalent impact on the electricity consumption. The analysis was re-run to evaluate the results obtained for the new "worst case" scenario profile. This profile is based on the island of Cyprus that in similitude with Sark has a high tourism arrival during summer months thus a higher energy consumption.

9. Further Studies and Suggestions

- Implement data research for the load consumption of the island. With an actual island energy profile, the project could change drastically due to the fact that the energy profiles and times of peak demands dictate which renewable energy is more suitable.
- Install a weather station to monitor the wind speed, solar irradiance and tidal current around the island. An analysis of tidal generation was outside the scope of this study, but with better access to data, this could be integrated into this feasibility study.
- The simplified economical assessment in the study was meant to showcase the difference possible outcome between mixes of different technologies. To have a more accurate economical assessment, an actual cost of the island energy generation will be required.
- Evaluate economic incentives for energy distribution to ensure a stable diesel generation output even when most of the energy is produced by renewables.
- Evaluate the interconnection of the island to produce a more distributed energy system that can stabilize the use of the renewables on the island.
- Include incentives for the installation of renewable energy via government programs.

Supplementary Materials: The following are available online at http://www.mdpi.com/1996-1073/12/24/4722/s1, Figure S1: UK-CY Load comparison, Figure S2: UK-CY Temperature comparison, Figure S3: Annual energy balance under different levels of renewable energy production, Figure S4: Case 1: Scenario 5 instantaneous generation and load variation (top) and energy balance (bottom), Figure S5: Case 1: Scenario 6 instantaneous generation and load variation (top) and energy balance (bottom), Table S1: Case 1: Summary of all energy mix scenarios with estimated battery size comparison with the two different load profiles, Figure S6: Case 1 installation and O & M, Table S2: UK-CY: Case 6 performance evaluation and comparison under different load profiles.

Author Contributions: Conceptualization, S.P. and T.M.; methodology, S.P., S.R. and T.M.; software, S.P., S.R. and T.M.; validation, S.P., S.R. and T.M.; formal analysis, S.P., S.R. and T.M.; investigation, S.P., S.R. and T.M.; resources, S.P., S.R. and T.M.; data curation, S.P., S.R. and T.M; writing—original draft preparation, E.J.G.; writing—review and editing, E.J.G.; visualization, E.J.G.; supervision, E.J.G.

Funding: This research received no external funding.

Conflicts of Interest: The authors declare no conflict of interest.

References

1. GOV.UK. 2019. Available online: https://www.gov.uk/guidance/2050-pathways-analysis (accessed on 22 July 2019).
2. Frey, B. *The SAGE Encyclopedia of Educational Research, Measurement, and Evaluation*; SAGE Publications, Inc.: Thousand Oaks, CA, USA, 2018; Volume 1–4. [CrossRef]
3. White, A. *Office of the Sark Electricity Price Control Commissioner Electricity Prices- Price Control Order*; Sark: London, UK, 2018.
4. Dalton, G.J.; Lockington, D.A.; Baldock, T.E. Case study feasibility analysis of renewable energy supply options for small to medium-sized tourist accommodations. *Renew. Energy* **2009**, *34*, 1134–1144. [CrossRef]
5. BBC NEWS. Sark Electricity: Deal Struck for Government Buy out. 2018. Available online: https://www.bbc.com/news/world-europe-guernsey-46397767 (accessed on 13 April 2019).
6. Nelson, J.; Gambhir, A.; Ekins-Daukes, N. *Solar Power for CO_2 Mitigation*; Briefing Paper 11; Imperial College London: London, UK; Grantham Institute for Climate Change: London, UK, 2014. Available online: https://www.imperial.ac.uk/media/imperial-college/grantham-institute/public/publications/briefing-papers/Solar-power-for-CO2-mitigation---Grantham-BP-11.pdf (accessed on 13 March 2019).
7. Ghenai, C. Life Cycle Analysis of Wind Turbine. In *Sustainable Development. Energy, Engineering and Technologies—Manufacturing and Environment*; Chaouki, G., Ed.; IntechOpen: London, UK, 2016; ISBN 978-953-51-0165-9. Available online: http://www.intechopen.com/books/sustainable-development-energy-engineering-andtechnologies-manufacturing-and-environment/life-cycle-analysis-of-wind-turbine (accessed on 4 February 2019).
8. Jovanovic, S.; Savic, S.; Bojic, M.; Djordjevic, Z.; Nikoli, D. The impact of the mean daily air temperature change on electricity consumption. *Energy* **2015**, *88*, 604–609. [CrossRef]
9. NASA. 2018. Available online: https://power.larc.nasa.gov/data-access-viewer/ (accessed on 13 November 2018).
10. Entsoe. Reliable Sustainable Connected. 2019. Available online: https://www.entsoe.eu/data/ (accessed on 13 April 2019).
11. Sark Electricity Limited. 2019. Available online: http://www.sarkelectricity.com/ (accessed on 14 April 2019).
12. Wind turbine models. 2019. Available online: https://en.wind-turbine-models.com/turbines/1533-atb-riva-calzoni-atb-500.54 (accessed on 19 January 2019).
13. McEvoy, A.; Markvart, T. *Solar Cells: Materials, Manufacture and Operation*, 2nd ed.; McEvoy, A., Castaner, L., Markvart, T., Eds.; Elsevier: Amsterdam, The Netherlands, 2012; p. 600.
14. Google. 2018. Available online: https://www.google.com/maps/place/55%C2%B057\T1\textquoteright00.0%22N+3%C2%B012\T1\textquoteright00.0%22W/@55.9494725,-3.2036368,14.5z/data=!4m5!3m4!1s0x0:0x0!8m2!3d55.95!4d-3.2 (accessed on 25 November 2018).
15. Muneer, T. *Solar Radiation and Daylight Models*, 2nd ed.; Elsevier: Amsterdam, The Netherlands, 2004.
16. ASHRAE. Refrigerating and Air-Conditioning Engineers. In *Handbook of Fundamentals*; American Society of Heating: New York, NY, USA, 1997.
17. GOV.UK. *Chapter 6 Renewable Sources of Energy*; GOV.UK: London, UK, 2018.
18. Autarsys Large ESS. 2017. Available online: https://www.autarsys.com/products/product/large-ess/ (accessed on 19 January 2019).
19. Narec Distributed Energy. *Sark Electricity Review–Island Grid with Wind, Solar and Batteries*; Office of the Sark Electricity Price Control Commissioner: London, UK, 2018. Available online: http://www.epc.sark.gg/assets/narec_island_system.pdf (accessed on 18 February 2019).

Article

Numerical Investigation on Heat Pipe Spanwise Spacing to Determine Optimum Configuration for Passive Cooling of Photovoltaic Panels

Samiya Aamir Al-Mabsali [1], Hassam Nasarullah Chaudhry [1,*] and Mehreen Saleem Gul [2,*]

1 School of Energy, Geoscience, Infrastructure and Society, Heriot-Watt University, P.O. Box 294 345, Dubai, UAE; saa83@hw.ac.uk
2 School of Energy, Geoscience, Infrastructure and Society, Heriot-Watt University, Edinburgh EH14 4AS, UK
* Correspondence: H.N.Chaudhry@hw.ac.uk (H.N.C.); M.Gul@hw.ac.uk (M.S.G.)

Received: 30 October 2019; Accepted: 5 December 2019; Published: 6 December 2019

Abstract: The uncertainty regarding the capacity of photovoltaics to generate adequate renewable power remains problematic due to very high temperatures in countries experiencing extreme climates. This study analyses the potential of heat pipes as a passive cooling mechanism for solar photovoltaic panels in the Ecohouse of the Higher Colleges of Technology, Oman, using computational fluid dynamics (CFD). A baseline model has been set-up comprised of 20 units, 20 mm diameter water-filled heat pipes, with a length of 992 mm attached to a photovoltaic panel measuring 1956 mm × 992 mm. Using the source temperature of 64.5 °C (337.65 K), the findings of this work have established that a temperature reduction in the range of up to 9 °C is achievable when integrating heat pipes into photovoltaic panels. An optimum spacing of 50 mm (2.5 times the diameter of the heat pipe) was determined through this work, which is also a proof-of-concept towards the use of heat pipe technology for passive cooling of photovoltaic panels in hot climates.

Keywords: CFD; heat pipe; temperature; photovoltaic; spanwise

1. Introduction

The potential of using photovoltaic (PV) panels operating in areas with hot and dry climates, such as Oman, is vast due to the abundance of solar radiation. However, despite this advantage, the uncertainty regarding PV panels to generate adequate renewable power is still a problem due to extreme temperatures. High temperatures lead to a reduction in the open circuit voltage of a PV system, thus lowering the power output [1–4]. According to Al-Waeli et al. [5], the PV cell temperature increase has two consequences: the reduction of the generated electrical energy and the thermal fatigue due to the significant temperature of the PV panel during the day.

Energy demand in Oman is on the rise, and the country has had a 5% energy consumption rise since 2015 according to Al-Mabsali et al. [6], who presented his work in the Oman Annual Report for 2016. The increase in energy demand is a challenge and the present study aims to optimise the renewable solar photovoltaic technology by increasing its operative range, which could contribute towards meeting the 5% energy consumption increase. The present study is also consistent with the vision of Higher Colleges of Technology (HCT) in the contribution to the national socio-economic development of the country through diversification to non-oil industries. Nasir and Al-Jabri [7] cited the Oman policy document "Vision 2020", which stated that industrial diversification can achieve the development targets, which included the increase in the contribution of the non-oil sectors and non-oil exports to 13% of the GDP by 2020.

In Muscat, the capital city of Oman, the hottest month is June where the average daytime temperature is around 40 °C (313.15 K), while in July and August, cloud banks brought by the

Southwest monsoon can slightly lower the daytime temperature to 38 °C (311.15 K) in July and to 36 °C (309.15 K) in August. In an earlier experiment carried out by the research team (Al-Mabsali et al.) [6], which was set up in the HCT Muscat Eco house, an average solar irradiation of 911 W/m^2 was observed in the months of June to July, 2017. Even during testing in a cooler period of the year such as October (testing period between 15th to 21st October 2017) (Table 1 and Figure 1), it was noted that the maximum nominal operating cell temperature (NOCT) was 64.5 °C (337.65 K) at an ambient temperature of 38.5 °C (311.65 K). The effect of this result was a decrease in performance efficiency of the PV panels by 2.19%. A maximum efficiency of 54.8% was recorded during the testing period, clearly indicating the adverse effect of hot arid climates on PV performance. This efficiency decline was in-line with the studies conducted by Al-Waeli et al. and Jouhara et al. [5,8] which show a significant drawback on the effectiveness of PV panels operating in hot arid climates.

Table 1. Experimental observation using data loggers.

Date	Ambient Temp. (°C)	NOCT (°C)	S (W/m^2)	Data Logger Temp. Readings		PV Cell Temp. (°C)	Actual Total Power Production (Watts)	Actual Power Production per Panel (Watts)	Rated Power Production per Panel (Watts)	Efficiency (%)
				Upper (°C)	Lower (°C)					
15/10/2017	38.5	64.5	911.11	64.5	46.9	89.18	12000	157.89	300	52.63
16/10/2017	38	63.8	911.11	63.8	47.8	87.88	12000	157.89	300	52.63
17/10/2017	36.5	61.4	911.11	61.4	48	83.65	12500	164.47	300	54.82
18/10/2017	33.7	59.2	911.11	59.2	44.9	78.34	12000	157.89	300	52.63
19/10/2017	32.2	57	911.11	57	42.8	74.34	12000	157.89	300	52.63
20/10/2017	32.3	58.5	911.11	58.5	43.3	76.15	9500	125.00	300	41.67
21/10/2017	31.5	57	911.11	57	44.9	73.64	8500	111.84	300	37.28

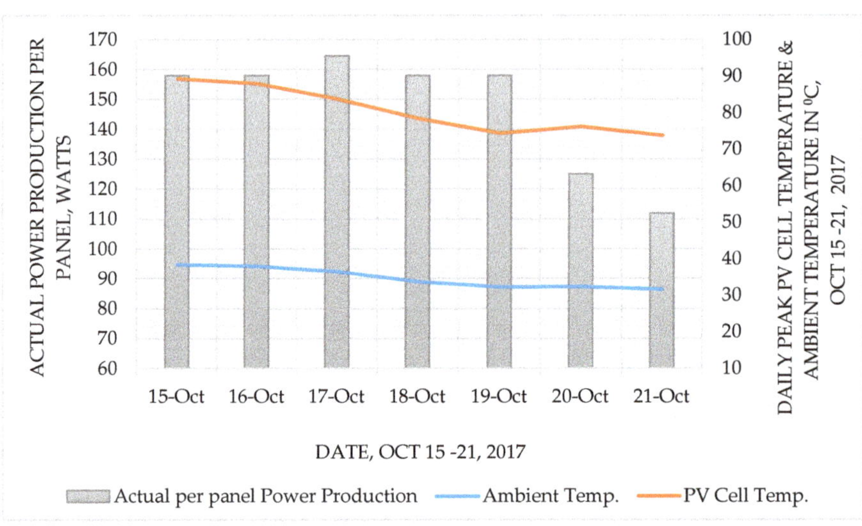

Figure 1. Actual photovoltaic (PV) power production at various daily peak PV cells and ambient temperatures.

In order to improve the power generation performance of PV panels, cell surface temperatures must be decreased to bring them closer to the ambient conditions; therefore, this study introduces a heat pipe heat exchanger (HPHE) technology as a passive cooling mechanism to be integrated within PV terminals.

Previous research carried out by Chaudhry [9] revealed that heat pipes incorporated with sorption phenomenon display greater heat transfer capacity and tubular heat pipes have the highest working range on average with the maximum operating temperature from all compared systems being 180 °C

(453 K). To maintain the sustainable working mode, it is imperative that heat pipes use water as a natural refrigerant in comparison to artificial refrigerants [9].

In this work, an investigation has been carried out using a heat pipe heat exchanger (HPHE) system as a retrofit mechanism for passively cooling PV panels. Computational fluid dynamics (CFD) is used to determine the optimum arrangement of heat pipes integrated with PV panels. The study uses the existing HCT Ecohouse as a case study where the PV panels are uniformly installed in a configuration of seventy-six (76) single PV panels of size 1956 × 992 mm, with a total PV array area of 147.44 m^2, delivering a direct current rating maximum power capacity of 22,800 Watts. The HPHE mechanism was arranged in a series of possible spanwise installations to discover the most functional design.

2. Literature Review

Previous studies have investigated various heat pipe systems for passive cooling duties in ventilation works and power electronics, which are summarized in this section. Peng et al. [10] investigated the practical effects of solar PV surface temperature efficiency based on its output performance. The experimental works were carried out under different radiation conditions and explored the variation of the output voltage, current, output power, and efficiency. The cooling test resulted in an efficiency increase of 47% for the PV panels. The system performance and life cycle assessment suggested that the annual PV electric output efficiencies could increase by up to 35%, and the annual total system energy efficiency, including electrical output and hot water energy output can increase by up to 107%.

Bahaidarah, Baloch, and Gandhidasan [11] carried out a review that highlighted the importance of uniform PV cooling. An experimental case study was presented for comparison between uniform and non-uniform cooling methods. The work explored and analyzed the possible causes and effects of non-uniformity using the cooling techniques with low average cell temperatures and uniform temperature distributions. One of these techniques was the utilization of heat pipes on PV systems that resulted in the reduction of the temperature down to 32 °C, with the best-case temperature non-uniformity of 3 °C.

Bahaidarah et al. [12] conducted a comprehensive study on the state-of-the-art applications, materials, performance of current heat pipe devices and future developments in the field, the current limitations of heat pipes, and the reasons that it cannot be implemented in more aspects of our lives due to its operational boundaries, cost concerns, and the lack of detailed theoretical and simulation analysis. The limitations resulting from their review provided the opportunity to find fresh solutions which opened up the possibilities of adopting the heat pipe technology to its feasible and fruitful utilization and thus used as the basis to achieve the objective of this study.

Jouhara et al. [13] experimented with PV efficiency caused by the water cooling effect using a numerical model, EES (Engineering Equation Solver), which predicted electrical and thermal parameters affecting its performance. A heat exchanger, as a cooling panel, was incorporated in the rear surface of the PV module as an experiment. The results of the numerical model were found in good agreement with the experimental measurements performed for the climate of Dhahran, Saudi Arabia. With active water cooling, the module temperature dropped significantly to about 20% and that increased the PV panel efficiency by 9%.

Theoretical modelling was carried out by Chaudhry et al. [14] using water as the liquid medium for ventilation works and utilizing a water flow rate of 0.25 m/s. The study used various spacing from 1D to 4D with a maximum of 80 mm down to 20 mm. The spacing range consideration of 1D to 2D, 20mm to 40 mm was subject to an international patent application (PCT/GB2014/052263). The 2D to 4D, 40 to 80 mm was a novel observation applied to a heat pipe installed in a duct as PV cooling device. Furthermore, citing Reference [9], findings on a systematic design of a high conductivity cooling system revealed that heat pipes incorporated with sorption phenomenon displayed greater heat transfer capacity, and tubular heat pipes arrangements have the highest working range on average, with the maximum operating temperature from all compared systems being 180 °C (453 K).

Tripathy et al. [15] conducted a study on building integrated photovoltaic (BIPV) thermal technology using an air duct, provided below, on the PV panels to serve as a structural element. The contribution of the air flow was to increase both the electrical and thermal efficiencies. The study utilized the energy equilibrium equation for developing the mathematical model of BIPV thermal system using the HDKR (Hay, Davies, Klucher, Reindl) model based on insolation, corresponding to the optimum tilt angle of the panel. The room temperature of the BIPV thermal system had a mass flow rate of 1 kg/s through the duct on the respective optimum tilt angle.

Further investigation by Chaudhry et al. [14] included optimization of the heat pipe arrangement for natural ventilation using CFD and the wind tunnel method. The airflow and temperature profiles were numerically predicted, the findings of which were quantitatively validated using wind tunnel experimentation. Using a source temperature of 41 °C and an inlet velocity of 2.3 m/s, the stream wise distance-to-pipe diameter ratio varied from 1.0 to 2.0 and the emergent cooling capacities were established to comprehend the optimum arrangement. The results of this investigation indicated that the heat pipes operated at their maximum efficiency when the streamwise distance was identical to the diameter of the pipe as this formation allowed for the incoming airstream to achieve the maximum contact time with the surface of the pipes. The technology presented was subjected to an international patent application (PCT/GB2014/052263). Therefore, Chaudhry et al. [14] works were used as a benchmark methodology for the current study.

3. Research Methodology

Computational fluid dynamics (CFD) was used as the primary research method in this study by modelling the entire heat pipe integrated PV panel, made up of 20 units of HPHE installed below the 1956 × 992 mm PV surface. Three models of heat pipe arrangement within the HPHE were simulated and spaced at 60, 50, and 40 mm apart at a spanwise distance measured equally between the center of the heat pipes. A flow rate of 0.25 m/s was considered in the determination of the optimum configuration of the heat pipe. ANSYS Fluent (v14.5, ANSYS, Canonsburg, PA, USA) was used to perform the numerical simulations. Furthermore, the HPHE models were tested using both the single-sided and the double-sided condenser direction and were evaluated by shifting the locations from the top, middle, and bottom sections of the rectangular duct.

3.1. Physical Domain

For the heat pipe physical domain, two spacing methodologies were considered: streamwise and spanwise. The design installation in the streamwise direction was not efficient because of the long evaporator pipe length, which led to space restriction on the site. The spanwise heat pipe installation was selected to be studied. The HPHE were installed inside a duct with the same length of the PV panel. There were two HPHE design installations, which were classified into the single side condenser direction shown in Figures 2 and 3, and double side condenser direction of the PV panel shown in Figures 4 and 5.

The review findings of Jouhara et al. and Tan and Zhang [12,16], which included controlled factors such as pipe diameter, pipe thickness, and liquid medium, were adopted in this study. The independent variables were identified as flow rate, heat pipe (HP) spacing, and design installations and are shown in Table 2. All heat pipe parameters, which include diameter, working fluid, and operating temperature, were taken from the previous works of Chaudhry et al. [9,14], as part of the research team's earlier works. The dependent variables were PV temperature and energy efficiency. The relationships of the variables such as pipe spacing to temperature, flow rate to temperature, and HP direction (spanwise and streamwise) to temperature were analyzed using CFD. The single side and double side arrangements (Figures 3–6) were made through the top, middle, and bottom installations of the PV panel. A duct, regardless of the material composition shown in Figures 3–6, was installed below the PV panel and collected the high temperature to be absorbed by the evaporator section of the HPHE and had a negligible effect on the temperature, as proven by the solar collectors of Amp et al. [17].

Table 2. Summary of the recommended heat pipe heat exchanger (HPHE) system specifications and liquid medium as controlled variables.

Specification	Value
No. of units	20
Spacing	2.0 D, 2.5 D, 3D
Pipe Material	Copper
Pipe Diameter	20 mm
Evaporator Length	992 mm
Condenser Length	992 mm
Total Length	1959 mm
Working Fluid	H_2O
Flow Rate	0.25 m/s
H2O Operating Temperature	218–453 K
Orientation	Span/Stream Wise—90°

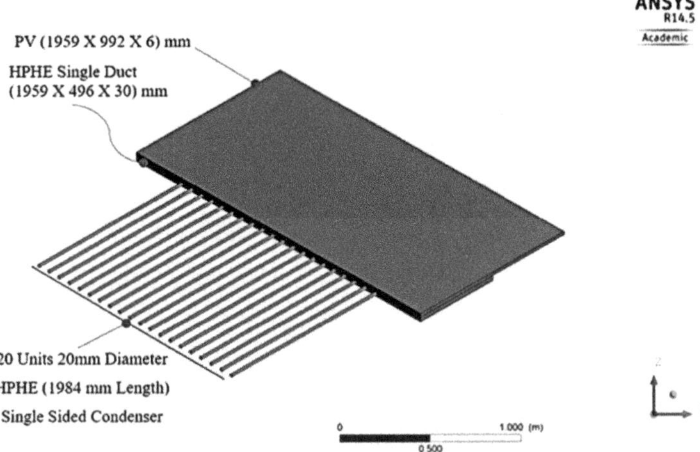

Figure 2. PV-HPHE spanwise with 50 mm spacing on the middle arrangement, single design isometric model.

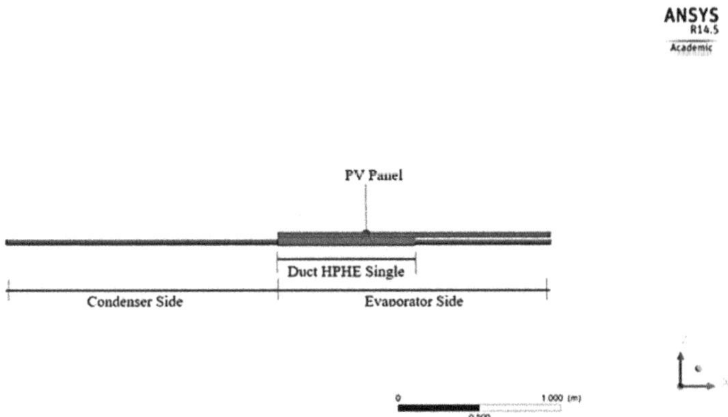

Figure 3. PV-HPHE spanwise with 50 mm spacing on the middle arrangement, single design side view model.

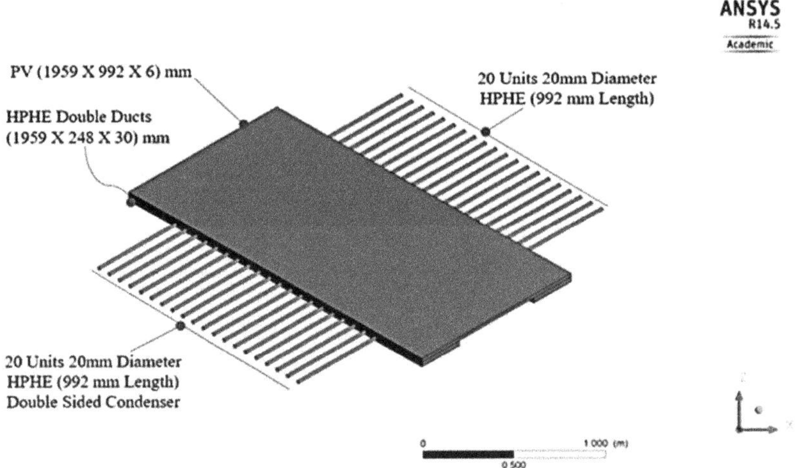

Figure 4. PV-HPHE spanwise with 50 mm spacing on the middle arrangement, double design isometric model.

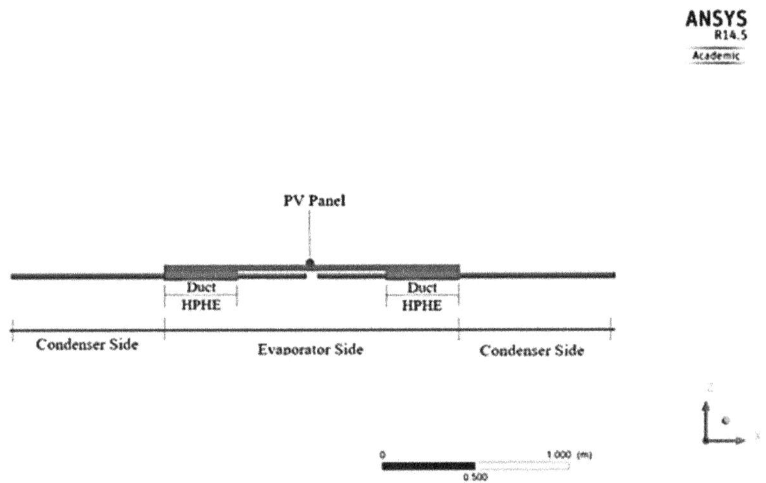

Figure 5. PV-HPHE spanwise with 50 mm spacing on the middle arrangement, double design side view model.

3.2. Mesh Generation and Boundary Conditions

A hexahedral mesh with nodes ranging between 703,507–1,135,505 and a maximum of 992,612 elements was applied to the different models in this simulation. The mesh orthogonal quality ranged from 0 to 1 with a minimum orthogonal quality equal to 7.12915×10^{-2} and a minimum aspect ratio equal to 9.59399×10 for a middle arrangement with spacing equal to 50 mm. The standard (*k-e*) *k*-epsilon turbulence model was applied with the standard wall function. In CFD, *k*-epsilon is a model used to simulate mean flow characteristics for turbulent flow conditions. The standard wall function was used to determine the laminar or turbulent flow of fluid material in the near wall using boundary conditions. Water was used as the working fluid inside the heat pipes. The mesh details are shown in Figure 6. The mesh quality results are shown below in Table 3.

Table 3. Mesh quality according to the heat pipe (HP) arrangement and spacing.

Mesh Quality	HP Arrangement	Heat Pipe (HP) Spacing (mm)						
		80	70	60	50	40	30	20
Nodes	HP Top	-	795712	748166	715756	726183	703507	716984
	HP Middle	1135505	864779	836655	788890	731874	704420	707050
	HP Bottom	-	825063	714460	727844	808531	754412	688898
Elements	HP Top	-	676937	634257	603092	614900	594312	604465
	HP Middle	992612	745573	722906	676522	621685	593749	598695
	HP Bottom	-	705884	601094	616130	697026	644068	688898

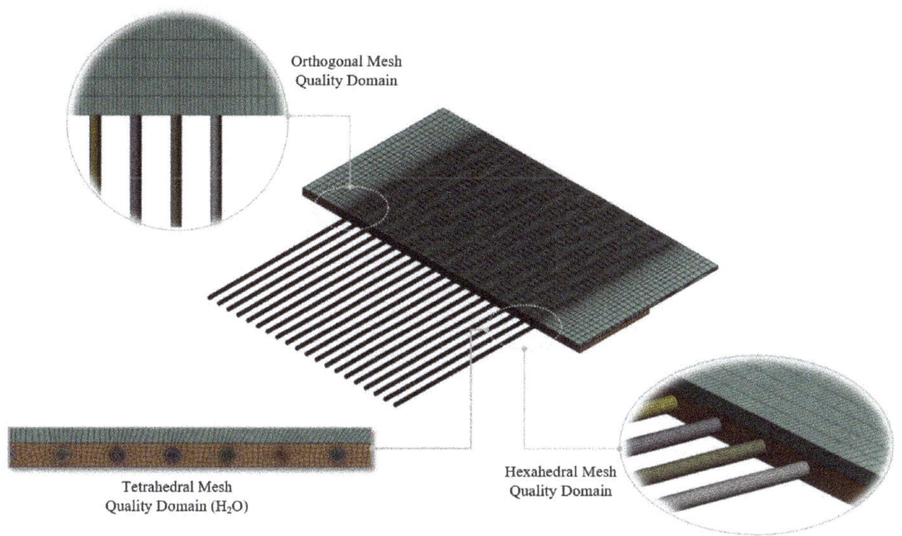

Figure 6. Mesh quality for the physical domain.

The solution method used a simple pressure velocity-coupling scheme. The spatial discretization was set to the following conditions: A gradient used the least square cell based, a standard pressure on fluid material, a momentum used second order upwind, a turbulent kinetic energy used first order upwind, a turbulent dissipation rate of first order upwind, and an energy was set using second order upwind. The reference values are shown in Table 4.

Table 4. Solution control variables.

Factors	Value	Units of Measurement
Area	1	m^2
Density	1.225	Kg/m^3
Enthalpy	0	j/Kg
Length	1	m
Pressure	0	Pa
Temperature	288.16	K
Velocity	0.25	m/s
Viscosity	1.7894×10^{-5}	Kg/(m·s)
Ratio of Specific Heats	1.4	-

The inlet temperature of liquid water in the HPHE was 45 °C (318.15 K). The top face of the PV panel models was assigned the source temperature of 64.5 °C (337.65 K) with an ambient temperature of 38.5 °C (311.65 K) taken from the July 2017 experimentally recorded data. The data revealed a decrease of 2.19% in performance efficiency of the PV panels, which was addressed by the passive cooling mechanism of the HPHE. This condition was in anticipation of the expected maximum surface temperature that was recorded during the experiments conducted earlier (presented in Table 1). The simulation of the flow of water within the pipe was done, the pipe inlet was assigned to the presumed evaporator end with the factors of mass flow rate set, and the range was 0.05 m/s as the minimum and 0.25 m/s as the maximum flow rate.

For simulation purposes, the temperature profile on the back surface of the PV panel was studied, and the results were translated to have a consequent effect affecting the top surface of the panel due to the thin layer of the PV panel. A simulation of the heat pipe liquid flow, with the use of CFD, ANSYS R14.5 was made. This approach allowed for relative properties, which have been applied as boundary conditions, as shown in Table 4. However, the HPHE system specifications and liquid medium, as controlled variables in Table 2, was focused on the decrease in temperature of working fluid, which stimulated the heat transfer from the photovoltaic panel to the evaporator section of the heat pipe, went through the condenser section, and achieved was through passive cooling. This allowed for a thorough observation of the passive cooling process of the HPHE design installation, which decreased the PV panel temperature from 2 to 5 °C. Additionally, for such an expected temperature reduction of the photovoltaic panel to have resulted, the consequent restoration in the loss of energy efficiency was 2.19%, which regained reliable power generation to the maximum available.

4. Results and Discussion

This section presents the findings from this work to determine the optimum heat pipe spanwise spacing using the numerical models shown in Section 3. Figure 7 displays the area weighted average temperatures on each of the simulated models. The minimum temperature in the HPHE evaporator section of 55.32 °C (328.47 K) was recorded on the HP middle model, which had a spacing of 50 mm (2.5 times the diameter of the HP or 2.5D). The highest temperature formation was observed for a heat pipe spacing of 40 mm or 2D in the HP top configuration. This result confirmed that having heat pipes spaced 2.5D apart from each other offers the highest passive cooling potential.

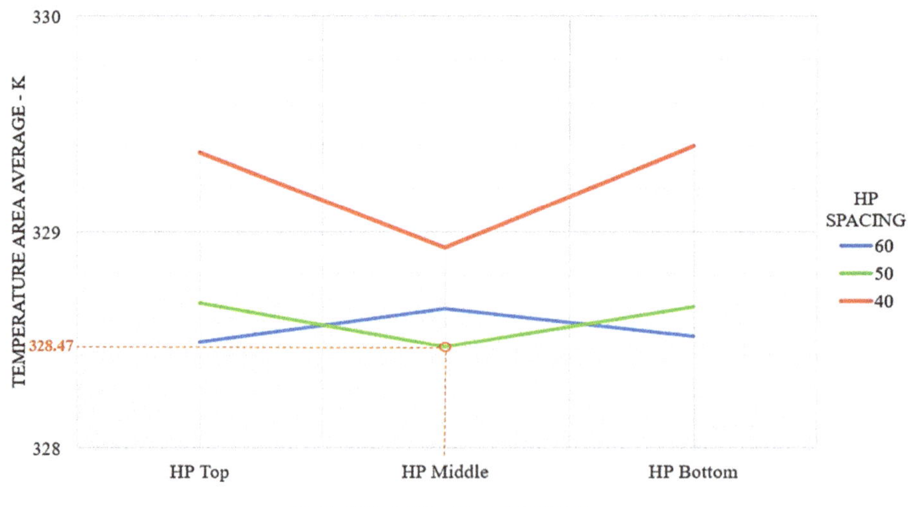

Figure 7. Temperature area average—K in the HPHE single duct.

The modelling set-up of the HPHE was done using a 40, 50, and 60 mm spacing on the centers with the same 20 units of HP installed in a rectangular duct, as shown in Figures 8–10. The installation caused the coverage of the cooling area of PV panel to slightly shrink, which resulted in varying results. For the 60 mm spacing, it resulted in 55.34 °C (328.49 K) at the top, 55.5 °C (328.65 K) in middle, and 55.37 °C (328.5 K) at the bottom, as shown in Figure 8.

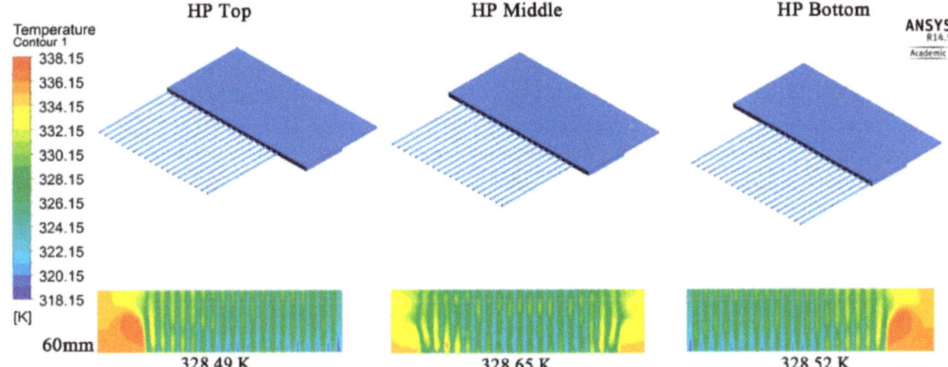

Figure 8. One sided condenser direction HPHE model, spacing = 60 mm, installation on PV panel.

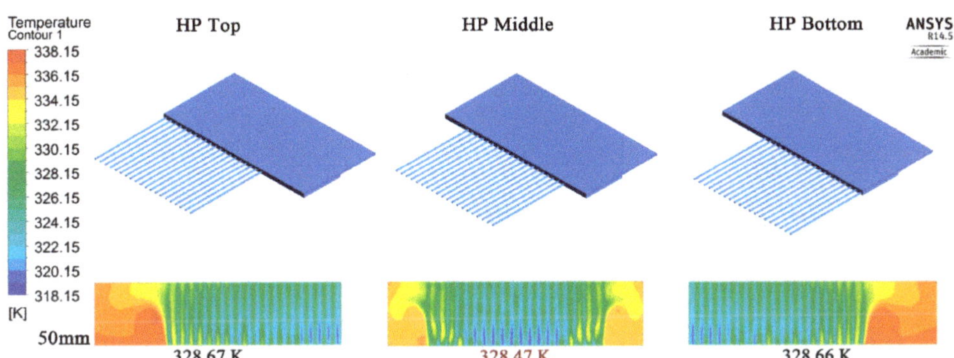

Figure 9. One sided condenser direction HPHE model, spacing = 50 mm, installation on PV panel.

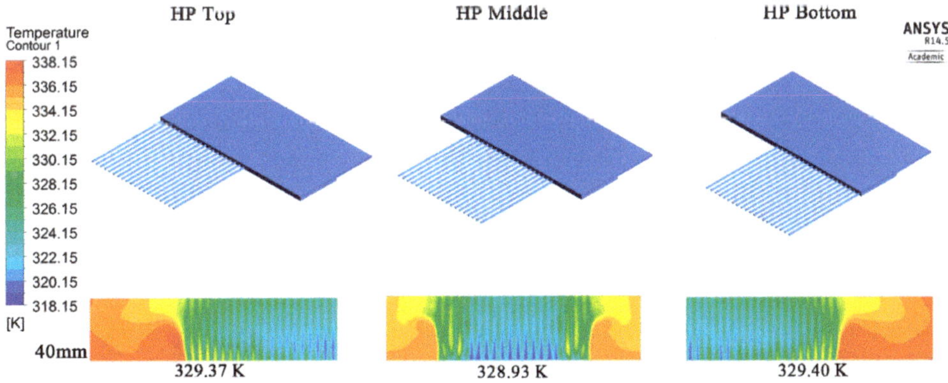

Figure 10. One sided condenser direction HPHE model, spacing = 40 mm, installation on PV panel.

The same HPHE installation with the use of 50 and 40 mm spacing's on the centers show 55.52 °C (328.67 K) at the top, 55.32 °C (328.47 K) in the middle, and 55.51 °C (328.66 K) at the bottom. For the 40 mm spacing 56.22 °C (329.37 K) at the top, 55.78 °C (328.93 K) in the middle, and 55.25 °C (328.40K) at the bottom were observed, as shown in Figures 9 and 10.

The three distinct HPHE design arrangements, which were made through the top, middle, and bottom installations of the PV panel, yielded theoretical results from ANSYS. The results showed that the average temperatures were 55.4 °C (328.55 K) for 60 mm HPHE spacing, 55.45 °C (328.60 K) for 50 mm HPHE spacing, and 56.08 °C (329.23 K) for 40 mm HPHE spacing. The results of the 50 mm HPHE spacing was selected for the reason that it yielded the lowest temperature of 55.32 °C (328.47 K).

The results of the 60 mm HPHE spacing gained the most uniform result. However, the problem was the middle section, which was the concentration of irradiation absorption for a long period of PV operation and experienced the highest temperature. The findings of the effect of air flow rate on the photovoltaic temperature is shown in Figure 11. Using variable values for the flow rate, which was ranged from 0.05 m/s to 0.25 m/s, the findings indicated an inverse relationship between the two parameters, as the heat pipe temperature decreased, the flow rate of the fluid increased. This was understandable because having a high airflow rate indicates a low contact time between the air and the heat pipe working fluid, thus providing lower cooling potential.

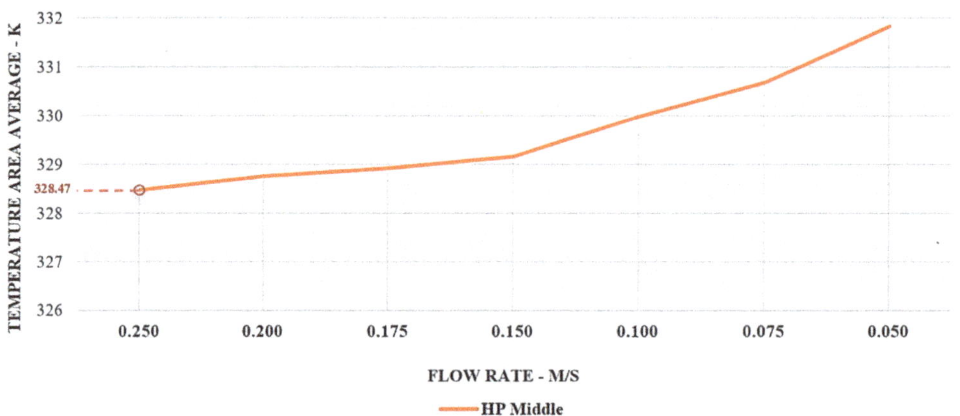

Figure 11. Variations of temperature results in relation to flow rate.

Temperature contour levels on the HPHE are shown in Figure 12, which were taken from the middle installation of the evaporator duct attached to the PV panel, as shown in Figures 4 and 5. The results of the optimum configuration yielded a minimum temperature of 54 °C (327.15 K) within the duct section.

The temperature of the heat pipe internal section inserted inside the duct of the HPHE was taken between the axial direction of 15 and 85, as shown in Figure 12. A maximum of 52.43 °C (325.58 K) and a minimum of 45.61 °C (318.76 K) were recorded as the range in temperature variations of the heat pipes that contained water and were inserted in the duct. The temperature was lower in relation to the duct section of the HPHE, which was 54 °C (327.15 K) and did not contain water. These results proved that the water inside the heat pipes acted as a conductor that transferred the warm temperature difference of 1.57 °C from the evaporator to the condenser section of the HPHE than the junction box, which caused the lowering of temperature below the PV panel, as shown in Figure 13.

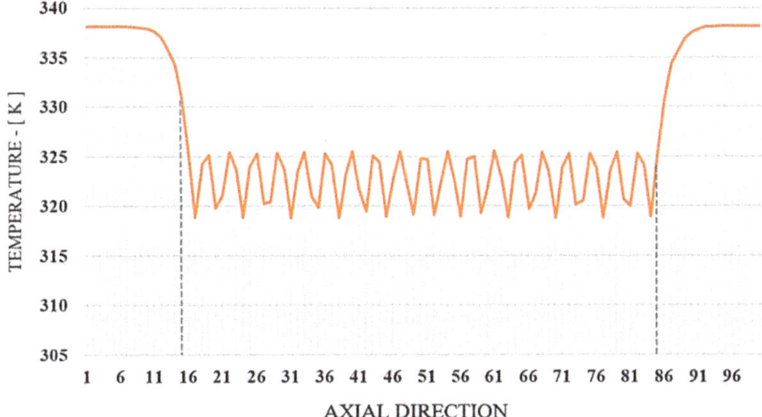

Figure 12. Temperature result of HPHE double sided condenser design.

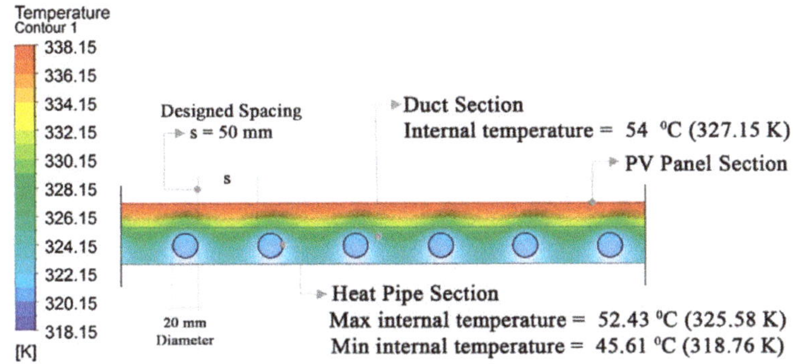

Figure 13. HPHE cross section with HP middle installation of the evaporator duct, spacing = 50 mm.

The findings from the three modelling set-ups made for the HPHE in the duct resulted in the selection of the best design and the conclusion is given in Table 5 below.

Table 5. Recommended HPHE design installation of the optimum configuration from CFD analysis.

Specification	Recommended Design
HPHE Arrangement	Double Sided Condenser Direction
HPHE Installation	Middle Section
Orientation	Span Wise—90°
Minimum Cooling Temperature	54 °C (327.15 K)
Axial Direction Range	15–85
No. of units	20
HPHE Duct Length	1959 mm
HPHE Duct Width	496 mm
HPHE Duct Thickness	30 mm
Pipes Spacing	2.5 D, s = 50 mm
Pipe Material	Copper
Pipe Diameter	20 mm

Table 5. *Cont.*

Specification	Recommended Design
Evaporator Length	496 mm
Condenser Length	496 mm
Working Fluid	H_2O
Flow Rate	0.25 m/s
H_2O Operating Temperature	218–453 K

As part of a qualitative visualization, the full-scale HPHE rig was fabricated using a double-sided condenser model, middle section installation, and 90 ° spanwise orientation, and is shown in Figures 14 and 15. This has been commissioned at the eco-house project location, at the HCT site in Muscat, Oman. The HPHE apparatus is currently undergoing experimental testing and data observation to validate the CFD results. All dimensions and specifications of the HPHE apparatus followed the experimental arrangement similar to the numerical model.

Figure 14. PV-HPHE apparatus isometric model (left) and captured infrared image (right) taken from the top surface assembled in the eco-house, Higher Colleges of Technology (HCT), Muscat, Oman.

Figure 15. PV-HPHE apparatus isometric model (left) and captured infrared image (right) taken from the bottom surface.

The commissioning of the apparatus shown in Figures 14 and 15 was carried out to experimentally validate the CFD results in order to prove the consistency of the modelling. The top surface temperature of PV-HPHE apparatus, last recorded 1st August 2019 at 12:52 h, was 69.84 °C and the bottom surface temperature was 64.98 °C. The difference in temperature between the top and bottom surface was 4.86 °C with the ambient temperature of 45.70 °C, which was in good agreement with the model. Overall, the simulation focused on heat transfer from the PV panel to the evaporator section of the heat pipe, through to the condenser section to exhaust the heat to the surroundings. The thorough observation of the passive cooling process of the HPHE design helped to decrease the PV panel temperature from 2 to 5 °C on average, which was confirmed from the actual installation.

5. Conclusions

The potential of PVs in hot and arid climates, although promising, may be adversely affected by the high intensity of solar radiation and high temperatures. This study carried out an investigation using a heat pipe heat exchanger (HPHE) system as a retrofit mechanism for passively cooling PV panels. Computational fluid dynamics (CFD) was used to determine the optimum spanwise arrangement of heat pipes integrated with PV panels.

The work undertaken analyzed the temperature formations on the PV panels using a range of heat pipe spacing combinations. The work has identified that the HPHE has the capacity to provide the required cooling of the photovoltaic panels installed in the HCT Ecohouse in Oman, which are usually exposed to a maximum temperature of 64.5 °C (337.65). The major finding from this study indicates that the 50 mm HPHE spacing (2.5D or 2.5 times the diameter of the pipe) has the greatest potential to decrease panel temperature, with a maximum reduction down to 55.32 °C (328.47 K) or approximately 9 °C. The recommended HPHE design installation is expected to be made of a double-sided condenser, having a middle section installation with a 90° spanwise orientation towards the PV panel. Current experimental testing has indicated a temperature drop between 2 to 5 °C, which is lower than the numerically predicted results.

This paper provided a proof-of-concept towards integrating heat pipes within PV panels to increase efficiency by 2.19% in order to restore the design power capacity specified in the previous design of the HCT Ecohouse, especially for PV panels operating in hot arid climates such as Oman. Furthermore, having heat pipes operating with water as the working fluid, as opposed to artificial refrigerants, underlines the suitability of this technology towards the development of sustainable solar energy in hot countries.

Author Contributions: Conceptualization, S.A.A.-M. and H.N.C.; methodology, S.A.A.-M., H.N.C. and M.S.G.; software, S.A.A.-M.; validation, S.A.A.-M.; formal analysis, S.A.A.-M.; investigation, H.N.C. and M.S.G.; resources, S.A.A.-M.; data curation, S.A.A.-M.; writing—original draft preparation, S.A.A.-M.; writing—review and editing, H.N.C. and M.S.G.; visualization, S.A.A.-M., H.N.C. and M.S.G.; supervision, H.N.C. and M.S.G.; project administration, H.N.C.; funding acquisition, S.A.A.-M.

Funding: This research received no external funding.

Acknowledgments: The authors would like to thank Heriot-Watt University and Higher Colleges of Technology, Oman for providing computational and experimental resources to carry out this research.

Conflicts of Interest: The authors declare no conflict of interest.

Nomenclature

T	Temperature
°C	Degrees Celsius
K	Temperature Kelvin
T_{cell}	PV cell temperature
T_{air}	Ambient temperature
Pa	Pressure in Pascal, N/m^2
m^2	Square meter
k-e	*k*-epsilon
Kg/m^3	Density, Kilogram per cubic meter
m/s	Meter per second
mm	Millimeter
NOCT	Nominal operating cell temperature
S	Insolation level = 1 W/m^2; ECO-House = 911 W/m^2
Sd/D	Streamwise distance-to-pipe diameter
D	Diameter, mm
kg/s	Kilogram per second
Watts/m^2	Watts per square meter
Kg/m-s	Viscosity, Kilogram per meter- second

Abbreviations

CFD	Computational Fluid Dynamics
PV	Photo Voltaic
EES	Engineering Equation Solver
HCT	Higher College of Technology
HPHE	Heat Pipe Heat Exchanger
HP	Heat Pipe

References

1. Kazem, H.A.; Chaichan, M.T. The effect of dust accumulation and cleaning methods on PV panels' outcomes based on an experimental study of six locations in Northern Oman. *Solar Energy* **2019**, *187*, 30–38. [CrossRef]
2. Gasparin, F.P.; Bühler, A.J.; Rampinelli, G.A.; Krenzinger, A. Statistical analysis of I–V curve parameters from photovoltaic modules. *Solar Energy* **2016**, *131*, 30–38. [CrossRef]
3. Kapsalis, V.; Karamanis, D. On the effect of roof added photovoltaics on building's energy demand. *Energy Build.* **2015**, *108*, 195–204. [CrossRef]
4. Al-Sabounchi, A.; Yalyali, S.A.; Al-Thani, H. Design and performance evaluation of a photovoltaic grid-connected system in hot weather conditions. *Renew. Energy* **2013**, *53*, 71–78. [CrossRef]
5. Al-Waeli, A.H.; Sopian, K.; Kazem, H.A.; Chaichan, M.T. Photovoltaic/Thermal (PV/T) systems: Status and future prospects. *Renew. Sustain. Energy Rev.* **2017**, *77* (Suppl. C), 109–130. [CrossRef]
6. Al-Mabsali, S.A.; Chaudhry, H.N.; Candido, J. Increasing the passive energy capacity of residential buildings with rooftop photovoltaic modules in hot and arid climates. In Proceedings of the 1st National Conference on Civil and Architectural Engineering, Sultan Qaboos University, Muscat, Oman, 26–28 March 2018.
7. Nasir, S.; Al-Jabri, K. Mineral Industry in Oman: In Country Value. In Proceedings of the 5th Fujairah International Industrial Rocks and Mining Forum, Fujairah, UAE, 18–20 April 2017.
8. Jouhara, H.; Milko, J.; Danielewicz, J.; Sayegh, M.A.; Szulgowska-Zgrzywa, M.; Ramos, J.B.; Lester, S.P. The performance of a novel flat heat pipe based thermal and PV/T (photovoltaic and thermal systems) solar collector that can be used as an energy-active building envelope material. *Energy* **2016**, *108* (Suppl. C), 148–154. [CrossRef]
9. Chaudhry, H.N. A study on optimising heat pipe geometrical parameters for sustainable passive cooling within the built environment. *Appl. Therm. Eng.* **2016**, *93*, 486–499. [CrossRef]
10. Peng, Z.; Herfatmanesh, M.R.; Liu, Y. Cooled solar PV panels for output energy efficiency optimisation. *Energy Convers. Manag.* **2017**, *150*, 949–955. [CrossRef]
11. Bahaidarah, H.M.S.; Baloch, A.A.B.; Gandhidasan, P. Uniform cooling of photovoltaic panels: A review. *Renew. Sustain. Energy Rev.* **2016**, *57*, 1520–1544. [CrossRef]
12. Jouhara, H.; Chauhan, A.; Nannou, T.; Almahmoud, S.; Delpech, B.; Wrobel, L.C. Heat pipe based systems—Advances and applications. *Energy* **2017**, *128*, 729. [CrossRef]
13. Bahaidarah, H.; Subhan, A.; Gandhidasan, P.; Rehman, S. Performance evaluation of a PV (photovoltaic) module by back surface water cooling for hot climatic conditions. *Energy* **2013**, *59*, 445–453. [CrossRef]
14. Chaudhry, H.N.; Calautit, J.K.; Hughes, B.R. Optimisation and analysis of a heat pipe assisted low-energy passive cooling system. *Energy Build.* **2017**, *143*, 220–233. [CrossRef]
15. Tripathy, M.; Yadav, S.; Panda, S.K.; Sadhu, P.K. Performance of building integrated photovoltaic thermal systems for the panels installed at optimum tilt angle. *Renew. Energy* **2017**, *113*, 1056–1069. [CrossRef]
16. Tan, R.; Zhang, Z. Heat pipe structure on heat transfer and energy saving performance of the wall implanted with heat pipes during the heating season. *Appl. Therm. Eng.* **2016**, *102*, 633–640. [CrossRef]
17. O'Hegarty, R.; Kinnane, O.; McCormack, S.J. Concrete solar collectors for façade integration: An experimental and numerical investigation. *Appl. Energy* **2017**, *206*, 1040–1061.

Article

Analysis of the Jeju Island Power System with an Offshore Wind Farm Applied to a Diode Rectifier HVDC

Sang Heon Chae [1], Min Hyeok Kang [1], Seung-Ho Song [2] and Eel-Hwan Kim [3,*]

1 Faculty of Applied Energy System, Jeju National University, Jeju 63243, Korea; chae@jejunu.ac.kr (S.H.C.); minh0131@jejunu.ac.kr (M.H.K.)
2 Department of Electrical Engineering, Kwangwoon University, Seoul 06978, Korea; ssh@kw.ac.kr
3 Department of Electrical Engineering, Jeju National University, Jeju 63243, Korea
* Correspondence: ehkim@jejunu.ac.kr; Tel.: +82-64-754-3674

Received: 21 October 2019; Accepted: 22 November 2019; Published: 27 November 2019

Abstract: The Jeju Island power system consists of two-unidirectional high voltage direct current transmission systems (HVDC), thermal power plants, and renewable energy sources. The local government's policy states that a 100 MW offshore wind farm should be constructed in the future. Due to the small size and sensitivity of the Jeju Island power system, power system analysis must be carried out before the installation of the new facility. Therefore, the objective of this study was to analyze the Jeju Island power system with a new wind farm applied to uncontrolled diode rectifier HVDC. Although there are many studies about the grid connection method of offshore wind farms, its small grid connection analysis has been rarely investigated, especially in the diode rectifier HVDC method. Diode rectifier HVDC is a new grid connection method for offshore wind farms, which reduces the costs and increases the reliability of the offshore platform. To verify the accuracy and effectiveness of simulation models, steady and transient state scenarios were conducted using the PSCAD/EMTDC program. First, the model of the Jeju Island power system without a new wind farm was compared with measured power system data. Second, its power system connected with a diode rectifier HVDC was simulated in a steady state. Finally, disconnection and single line ground fault occurred at the offshore wind farm, respectively. From the simulation results, the grid stability of the Jeju Island power system was confirmed considering a new facility.

Keywords: diode rectifier; HVDC; Jeju Island power system; offshore wind farm

1. Introduction

Jeju Island is one of the biggest islands in South Korea with a peak and average power demand of 944 and 627 MW, respectively, and consists of thermal power plants, renewable energy sources, two static synchronous compensators (STATCOMs) and two current source converter high voltage direct current (CSC-HVDC) transmission systems connected to the mainland. This small power system is being supplied with over 40% of the demand load by high voltage direct current transmission systems (HVDCs) from the power system on the mainland. However, the local government of this island has proceeded with the renewable energy promotion policy, namely "Carbon Free Island Jeju by 2030" [1]. Thus, an offshore wind farm (OWF) with a total capacity of 100 MW will be constructed in the near future in the north of Jeju Island. To achieve this plan successfully, the entire Jeju Island power system should be analyzed, including the large scale wind farm by using a detailed simulation model because this large scale wind farm will have a 16% average power load.

According to advanced research about HVDC with wind farm, voltage sourced HVDC can suitably deliver output power from wind farms to a weak grid such as the Jeju Island power system [2–7].

However, its application might increase the installation cost of an OWF. To deal with the economic challenge, [8–10] presented a new topology of HVDC, which changes conventional modular multilevel converter (MMC) to the uncontrolled diode rectifier at the offshore platform, as illustrated in Figure 1. Although the controllable converter will disappear in this new grid connection method, the diode rectifier HVDC (DR-HVDC) has many advantages, including reduced installation costs, low losses, easy management, and high reliability, among others.

From this perspective, this study analyzed the Jeju Island power system with a 100 MW OWF, which is connected to DR-HVDC via a 50 km submarine DC (Direct current) cable. To verify the effectiveness of DR-HVDC, a simulation model of the Jeju Island power system with a new OWF was conducted for steady and transient situations by using the PSCAD/EMTDC program. First, the Jeju Island power system, which was made by using actual parameters, operated without the new OWF. In this case, a comparison was made between the results of the simulated model and the measured data to check the accuracy of the simulation model. Second, a new OWF was linked to the Jeju Island power grid by DR-HVDC under normal operation. Third, the disconnection fault occurred to the OWF of the DC transmission line. Finally, the single line ground fault occurred to the OWF.

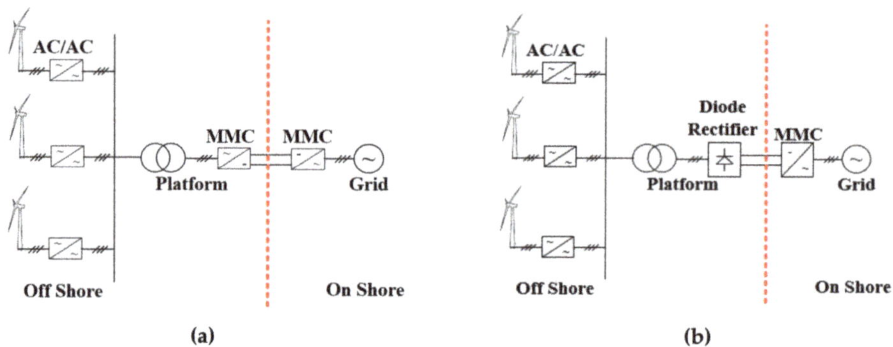

Figure 1. Conceptual design of high voltage direct current transmission systems (HVDC) for an offshore wind farm (OWF): (**a**) Conventional HVDC; (**b**) diode rectifier HVDC (DR-HVDC).

2. Modeling of the OWF System

2.1. Onshore MMC Station

An MMC is a type of voltage sourced converter (VSC), as shown in Figure 2. Using this concept, it is possible to make a huge capacity VSC [11]. In this case, the MMC plays a role as an onshore station of the new OWF by converting DC to AC (Alternative current), Figure 3. To transfer, MMC has three controllers, which are a current, a circulation current, and a capacitor balancing controllers [12–20]. Using the Park's transformation theory to control that, the terminal voltage of MMC in the dq axis can be calculated as

$$vtd = -Rid - pLid + vsd + \omega Liq \quad (1)$$

$$vtq = -Riq - pLiq + vsq - \omega Lid \quad (2)$$

where vt and vs are the terminal and grid voltage. i is the three-phase current. R and L are the resistance and inductance, respectively. p is the differential operator. ω is the grid angular frequency. If the PI controller is used, the current controllers will be expressed as

$$v^*td = -\ (Kp + Ki/s)\ (i^*d - id) + vsd + \omega Liq \quad (3)$$

$$v^*tq = -\ (Kp + Ki/s)\ (i^*q - iq) + vsq - \omega Lid \quad (4)$$

where superscript of * denotes reference mark. Then, i^*d and i^*q are decided by

$$i^*d = 2Q^*/ (3\ vsq) \tag{5}$$

$$i^*q = 2P^*/ (3\ vsq) + k\ (V^*dc - Vdc) \tag{6}$$

where P^* and Q^* are the real and reactive power, respectively. In this simulation, P^* will be the summation of the generated power from the OWF. Q^* will be zero to make the unity power factor. k is the coefficient for the *dc* link voltage control to maintain a stable range of it, as illustrated in Figure 3.

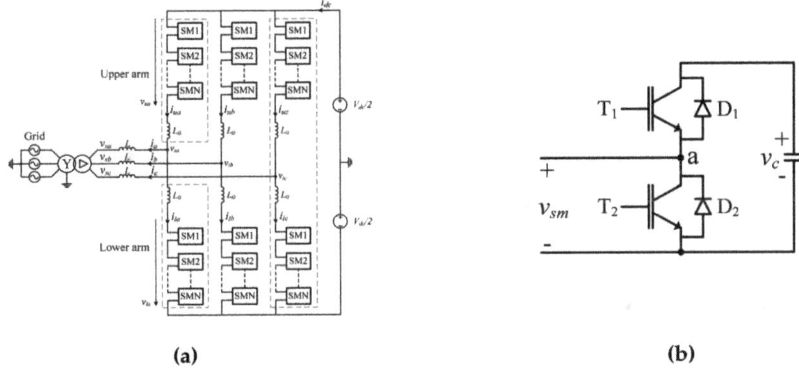

Figure 2. Basic structure of a modular multilevel converter (MMC): (**a**) Topology; (**b**) Submodule.

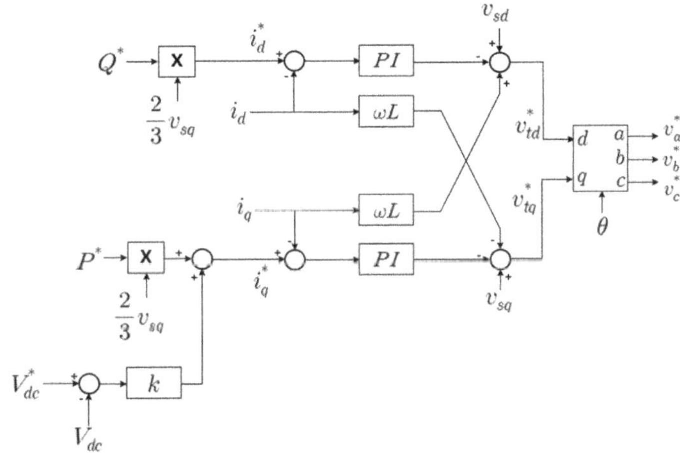

Figure 3. Current controller of MMC in an onshore station.

The MMC needs a circulation current controller to suppress it because it always occurs from the difference of capacitor voltages among phases. To mitigate circulation current, the differential voltage of the MMC in the dq frame can be written as

$$vdiffd = R0icird + pL0icird - 2\omega L0icirq \tag{7}$$

$$vdiffq = R0icirq - pL0icirq + 2\omega L0icird \tag{8}$$

where *vdiff* is the differential voltage. *R0* and *L0* are the resistance and inductance of arm inductors, respectively. *icir* is the circulating current. If the PI controller is adapted to the circulating current controller, as shown in Figure 4, it will be expressed as

$$v^*diffd = -(Kp + Ki/s)(i^*cird - icird) - 2\omega L0icirq \tag{9}$$

$$v^*diffq = -(Kp + Ki/s)(i^*cirq - icirq) + 2\omega L0icird \tag{10}$$

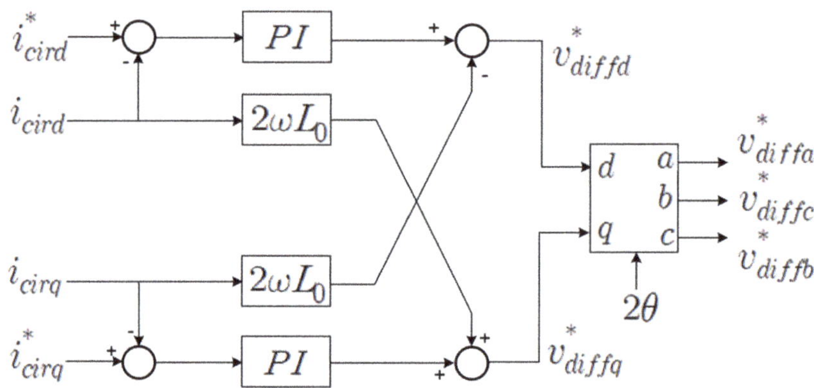

Figure 4. Circulating current controller of MMC in an onshore station.

The final reference value will be generated as a PWM (Pulse width modulation) switching signal, then it will be decided by a capacitor balancing controller depending on sorted capacitor voltages. The parameters of the MMC are as described in Table 1.

Table 1. Parameters of an MMC.

Quantity	Value
Active power	100 MW
Reactive power	50 MVar
Rated AC voltage	100 kV
Rated DC link voltage	200 kV
Grid frequency	60 Hz
Arm inductance	3.4 mH
Submodule Capacitance	7800 uF
Number of submodules per arm	20 EA
PWM method	Phase shift PWM

2.2. Offshore Diode Rectifier Station

To convert AC power from the wind power generator to DC power, an offshore DR station should be connected to the DC link of the onshore MMC station, as shown in Figure 5. It consists of a DR, AC filter, and phase-shifting transformers. The phase-shifting transformer can reduce the ripple voltage of the DC link. Through a series connection with them, the DC voltage will be increased to rated voltage. The parameters of the DR station are shown in Table 2.

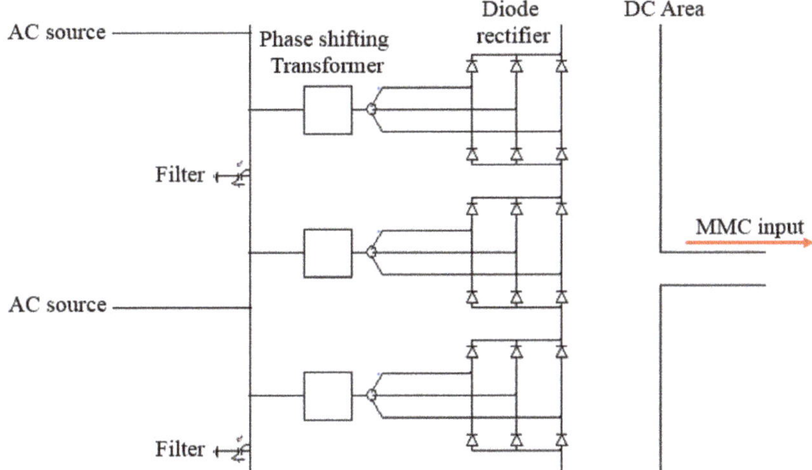

Figure 5. Simulation model of diode rectifier (DR) (Offshore side rectifier).

Table 2. Parameters of diode rectifier (DR) station.

Quantity	Value
Rated DC voltage	200 kV
Number of phase-shifting transformers	12 (72 pulses)
Number of 3-phase DRs	12
Length of DC link	50 km

2.3. Wind Turbine

In this analysis case, the simulation model of the offshore wind farm will be used as equivalent models to simplify the simulation model, i.e., 20 MW 2 level VSC, as illustrated in Figure 6, is assumed as four wind turbines each with a capacity of 5 MW. Although the wind turbines are replaced with the equivalent models, its controller will be quite similar to the detailed model, i.e., the current controller, which is similar to the MMC model, will be applied to the equivalent model. There is one problem of the controller in a wind turbine because the uncontrolled DR station cannot generate reference voltage signals and phase angle. It means that it is impossible to perform voltage transforming to dq from a 3-phase frame from the DR side converter of a wind turbine. Thus, [8–10] proposed the method as known as FixRef control, as illustrated in Figure 7, which uses GPS signal instead of space phasor angle of grid voltage. This study will also use a steady increased time signal, which is an assumed GPS signal.

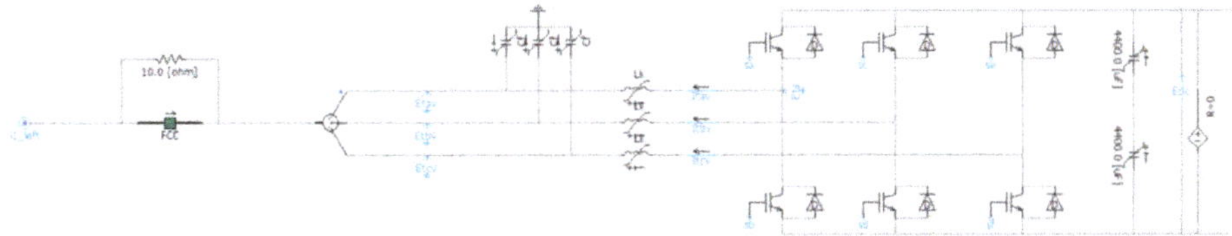

Figure 6. Equivalent simulation model of wind turbines in the new OWF.

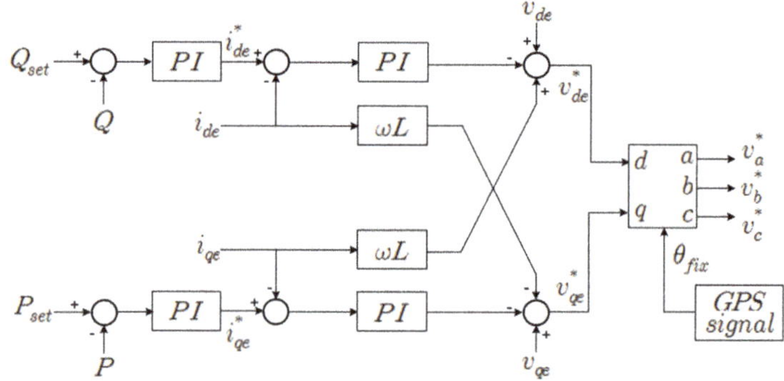

Figure 7. Equivalent simulation model of a wind turbine in the new OWF.

2.4. Whole OWF System

Figure 8 shows the whole simulation model of OWF with DR-HVDC. This system will be attached to the Jeju Island power system as the new OWF, then it will be simulated by parallel computing method in PSCAD/EMTDC program.

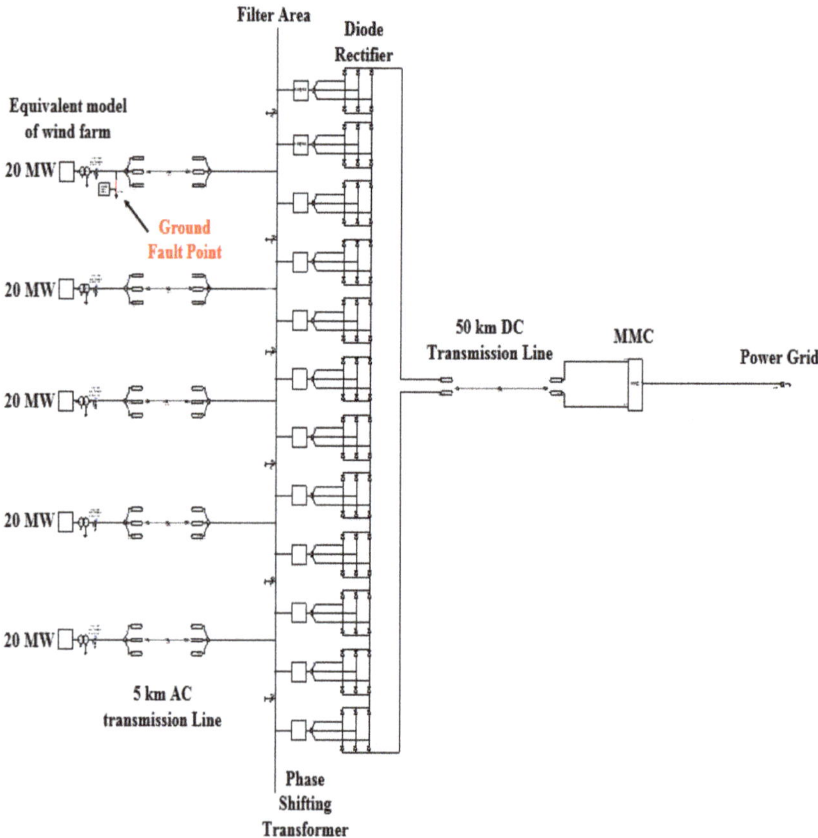

Figure 8. Simulation model of OWF with DR-HVDC.

3. Configuration and Modeling of the Jeju Power System

3.1. CSC-HVDC

In the Jeju Island Power system, two CSC-HVDC can operate only as unidirectional. One of which is a frequency regulator, and the other is supplying constant power. Figure 9 represents the simulation model of the CSC-HVDC, which consists of thyristors, a passive filter, a synchronous compensator, and its controller.

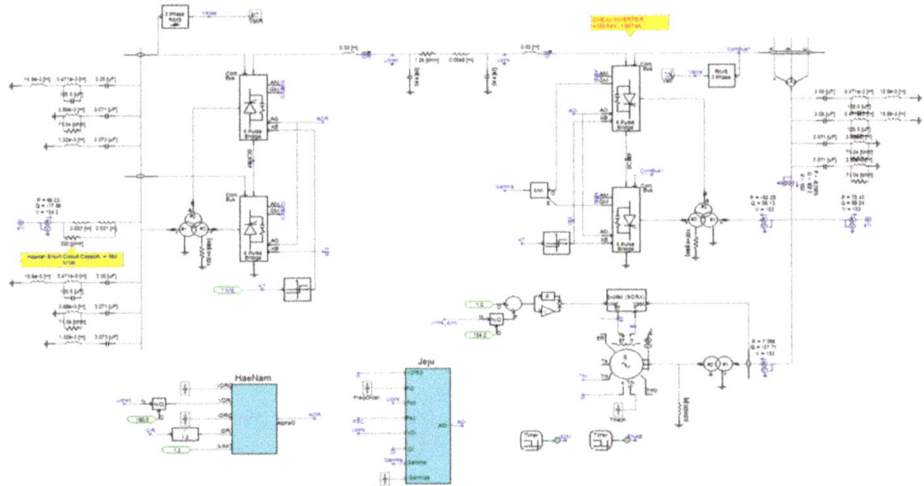

Figure 9. Simulation model of a current source converter high voltage direct current (CSC-HVDC) transmission system.

3.2. Thermal Power Plant

Three thermal power plants in the Jeju Island Power system are also operated separately. In this simulation case, the simplified equivalent controlled current sources are applied, as seen in Figure 10. The measured data will be used to operate this model.

3.3. STATCOM

The Jeju Island power system has two STATCOM, each with a capacity of 50 MVar. They have an important role with respect to grid voltage stability in its power system because the CSC-HVDCs consume a lot of reactive power. The simulation models of STATCOM consist of a three-level VSC and its controller, as shown in Figure 11.

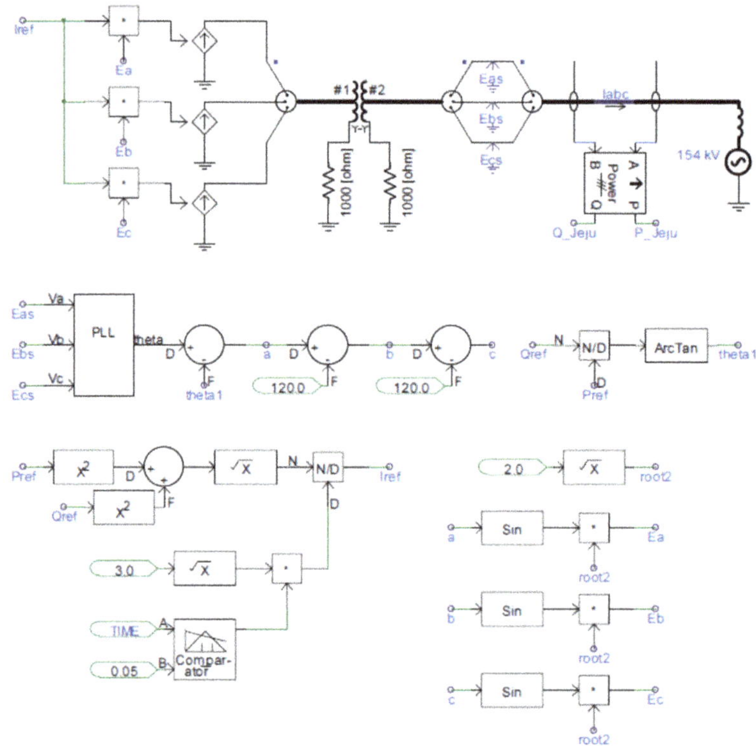

Figure 10. Simulation model of a thermal power plant.

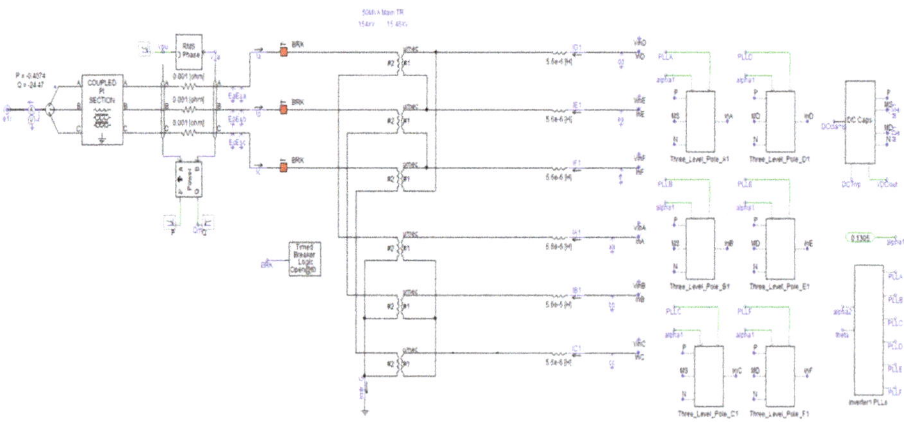

Figure 11. Simulation model of STATCOM.

3.4. Existing Wind Farms

The simulation models of existing wind farms with approximately 250 MW are used as controlled current source equivalent model, which can adjust active and reactive powers easily in a simulation program. Thus, the measured data in the Jeju Island power system will be input data of this model, as seen in Figure 12.

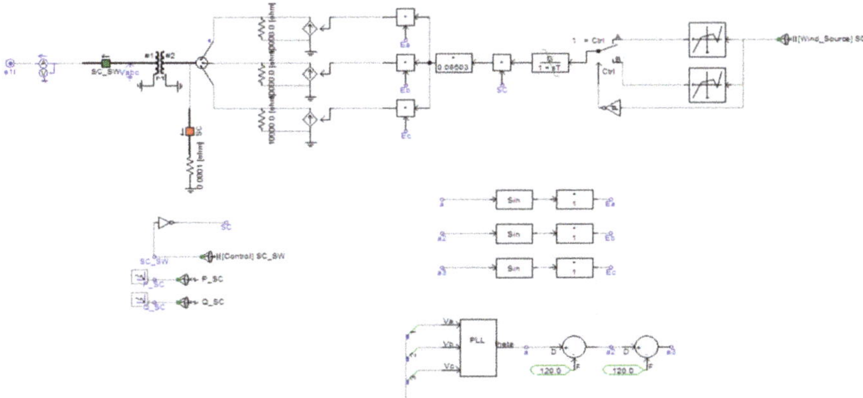

Figure 12. Simulation model of an existing wind farm.

3.5. Whole Simulation Model Including the New OWF

Figure 13 shows the whole simulation model of the Jeju Island power system. The red line represents the connecting point of the new OWF by an electric network interface (ENI), as located in the top left of Figure 13, which supports the parallel computing part in PSCAD/EMTDC. The time scale of the whole simulation is assumed to be 0.14 milli-times by adjusting the time constant of every component. The actual parameters and configurations of the transmission line are applied to this system.

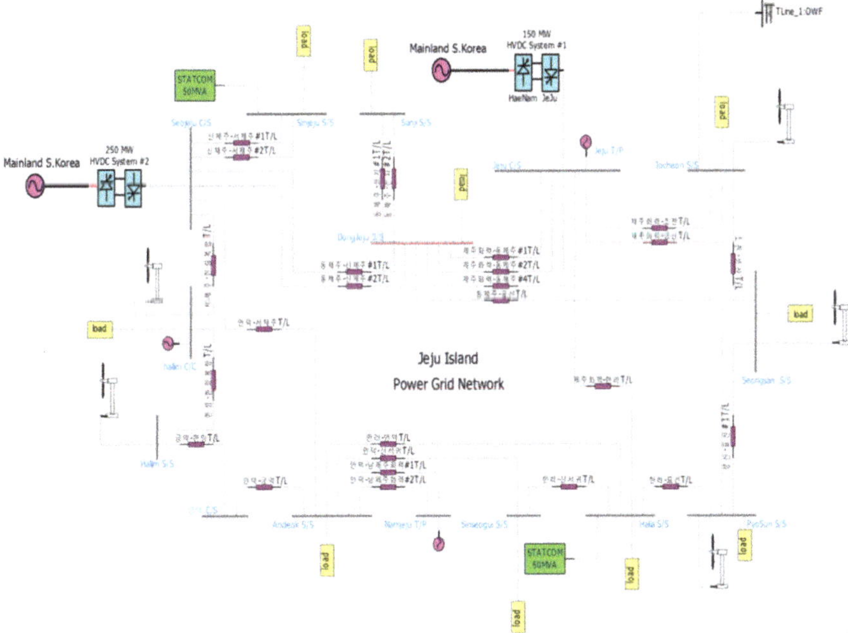

Figure 13. Simulation model of overall Jeju Island power system.

4. Simulation Results

Because there are many components of the Jeju Island power system, the line colors should be noted as the following colors in Table 3. This line color will be applied to every simulation result of active and reactive power.

Table 3. Expressions of simulation results.

Items	Line Color
Power load	Red
HVDC #1	Yellow
HVDC #2	Dark brown
Thermal power plants	Green
Existing wind farms	Blue
STATCOM #1	Grey
STATCOM #2	Pink
DR-HVDC (OWF)	Brown

4.1. Case 1: Normal Operation of the Jeju Power System without 100 MW OWF

To confirm the accuracy of the base simulation model, this scenario was conducted. Figure 14, representing active and reactive power at the top and bottom, respectively, shows measured data and simulation results. The top and bottom of Figure 15 represent the grid frequency and voltage. Hence, the errors between the simulation results and the measured data were less than 1% by grid frequency and voltage.

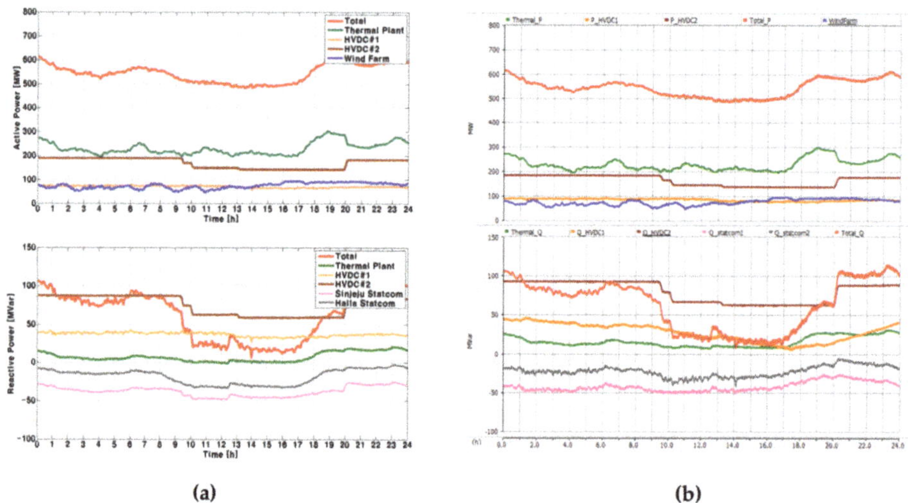

Figure 14. The simulation results of case 1: (**a**) Measured data in the Jeju Island power system (top: Active power, bottom: Reactive power); (**b**) The simulation model (top: Active power, bottom: Reactive power).

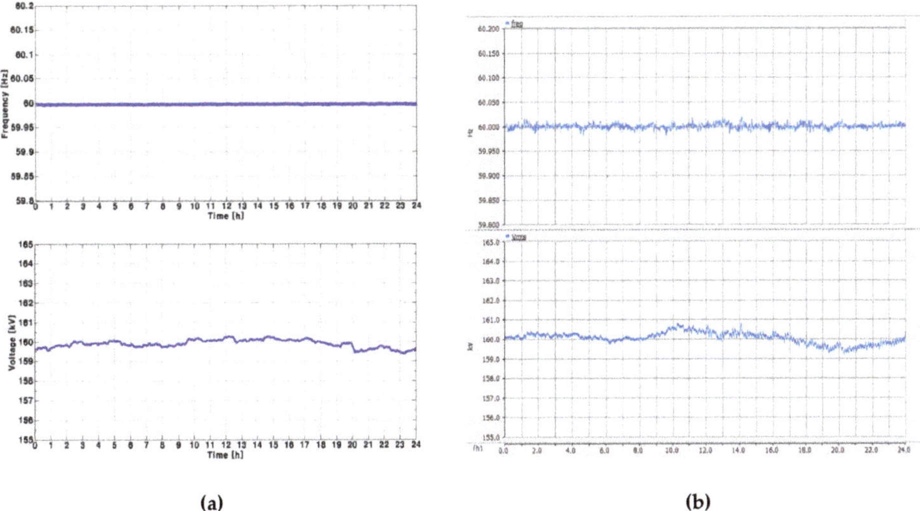

Figure 15. The simulation results of case 1: (**a**) Measured data in the Jeju Island power system (top: Grid frequency, bottom: Grid voltage measured at the biggest power load); (**b**) The simulation model (top: Grid frequency, bottom: Grid voltage measured at the biggest power load).

4.2. Case 2: Normal Operation of the Jeju Power System with a New 100 MW OWF Based on DR-HVDC

In the second scenario, the new 100 MW OWF was connected to the Jeju Island power system newly. In contrast to the first scenario, the CSC-HVDC #2 was operated at a limited minimum power, and the thermal power plants also reduced the output power, because the new OWF had generated additional active power. The CSC-HVDC #1 adjusted output power following demand power load and output from the wind farm to stabilize the grid frequency, as seen in Figure 16a. Due to the operation of CSC-HVDC #1, the grid frequency was in the grid code of South Korea from 59.8 Hz to 60.2 Hz. The maximum variance of frequency was slightly increased to 60.02 Hz, as compared to the first case. The variance of voltage was also higher in case 2 than in case 1 without the operation of the new OWF by approximately 2 kV, as shown in Figure 16b.

Figure 17a shows simulation results focused on DR-HVDC. Its reactive power was maintained to the unity power factor. The DC link voltage of DR-HVDC was in a constant value of 200 kV regardless of the active and reactive power, as seen in Figure 17b.

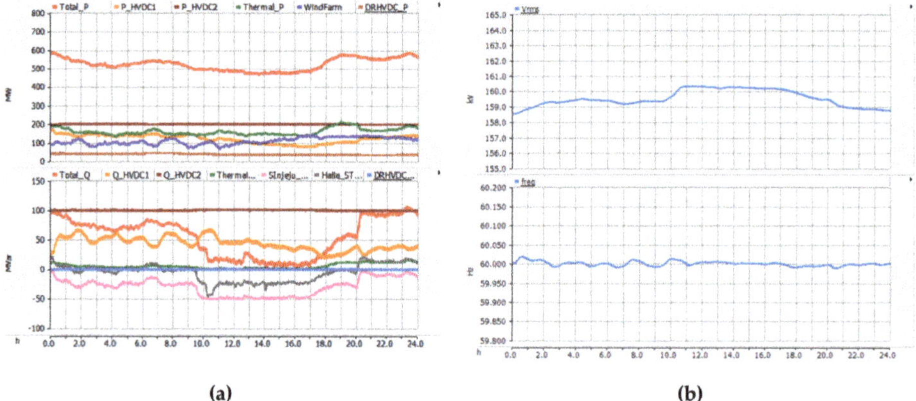

Figure 16. The simulation results of case 2: (**a**) top: Active power, bottom: Reactive power; (**b**) top: Grid frequency, bottom: Grid voltage measured at the biggest power load.

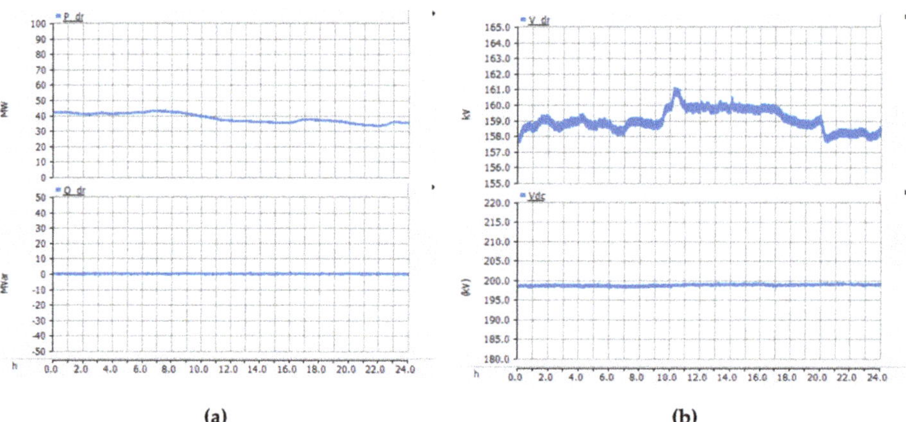

Figure 17. The simulation results of case 2: (**a**) top: Active power of DR HVDC, bottom: Reactive power of DR HVDC; (**b**) top: AC voltage measured at DR HVDC connection point, bottom: DC link voltage of DR HVDC.

4.3. Case 3: Disconnection Fault Occurred at the DC Transmission Line

In the third scenario, the DC submarine cable was disconnected abruptly. Thus, the output power of DR-HVDC was zero suddenly, and then the HVDC #1 compensated, as shown in Figure 18a. From this fault condition, the frequency and voltage were dropped to 59.97 Hz and 158 kV, as seen in Figure 18b, respectively. The DC link voltage of DR-HVDC also had a 5% variance. The AC voltage was recorded at the OWF grid by converter operation of wind turbines, as shown in Figure 19a. Figure 19b presents the onshore output of the DR-HVDC, whose active power was reduced to zero because of disconnection fault.

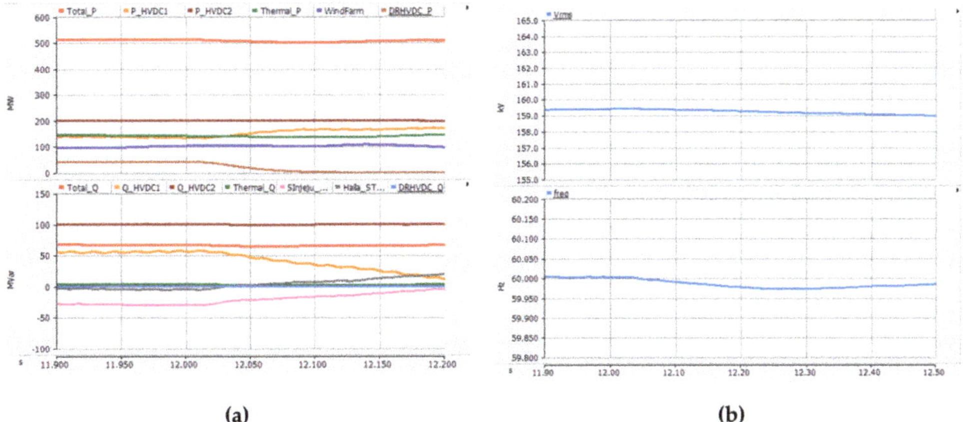

(a) (b)

Figure 18. The simulation results of case 4: (**a**) top: Active power, bottom: Reactive power; (**b**) top: Grid frequency, bottom: Grid voltage measured at the biggest power load.

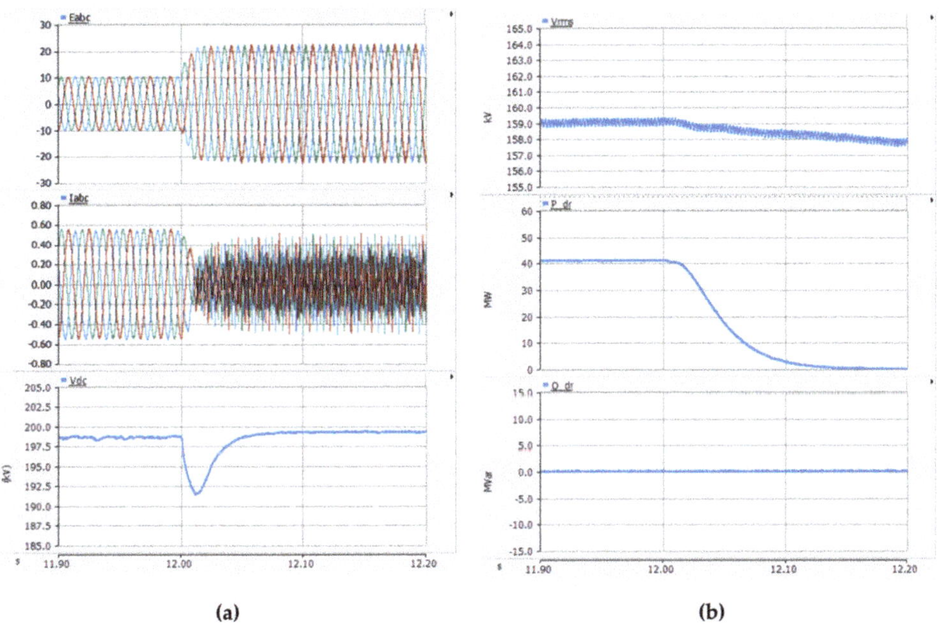

(a) (b)

Figure 19. The simulation results of case 4: (**a**) top: Instantaneous voltage measure at the OWF grid, middle: Instantaneous current measure at the OWF grid, bottom: DC link voltage of DR-HVDC; (**b**) top: AC voltage measured at DR HVDC connection point, middle: Active power of DC-HVDC, bottom: Reactive power of DR-HVDC.

4.4. Case 4: Single Line Ground Fault Occurred at AC Line in the New 100 MW OWF

From the simulation results of case 4, the DR-HVDC was able to protect the Jeju Island power system from an OWF side fault. Although the voltage and current of the OWF side power grid were oscillated, as shown in Figure 21a, the voltage and frequency of the Jeju Island power system was stable, as seen in Figure 20b. The active power dropped at 35 MW, as illustrated in Figure 21b, then the CSC-HVDC #1 compensated that immediately, as seen in Figure 20a.

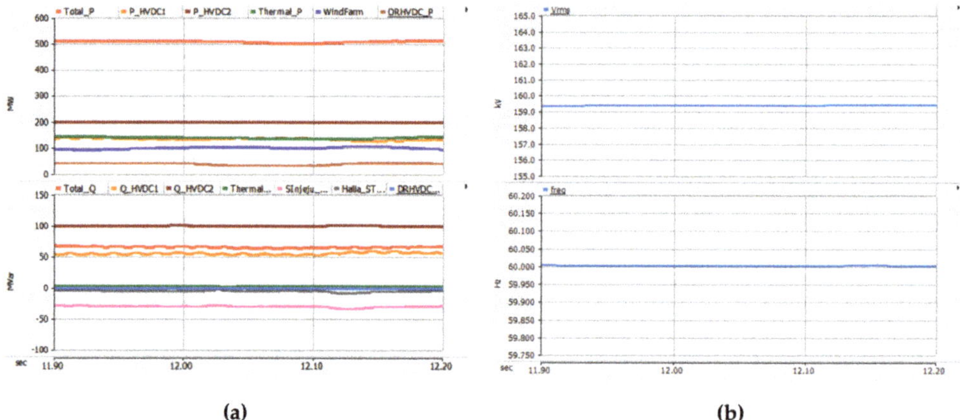

Figure 20. The simulation results of case 4: (**a**) top: Active power, bottom: Reactive power; (**b**) top: Grid frequency, bottom: Grid voltage measured at the biggest power load.

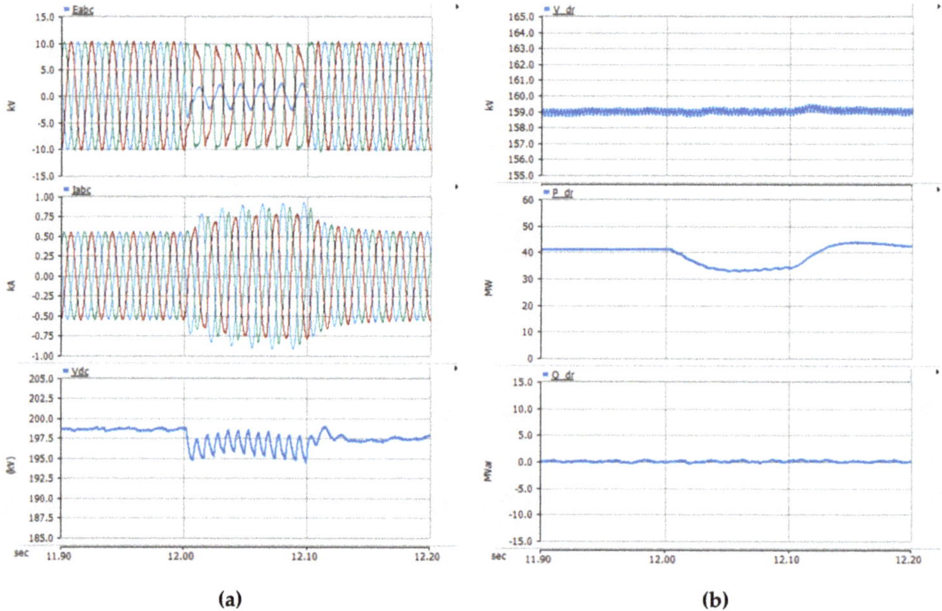

Figure 21. The simulation results of case 4: (**a**) top: Instantaneous voltage measure at OWF grid, middle: Instantaneous current measure at OWF grid, bottom: DC link voltage of DR-HVDC; (**b**) top: AC voltage measured at DR HVDC connection point, middle: Active power of DC-HVDC, bottom: Reactive power of DR-HVDC.

5. Conclusions

This study proposed installation of a DR-HVDC for the new 100 OWF in the Jeju Island power system. To confirm the impact of the DR-HVDC, the simulation model of the present Jeju Island power system was conducted and compared with the actual power system operating history in the first case. From the analysis results of the second case, its power system had been linked with the OWF by using the DR-HVDC. And it was able to operate a steady state when the DR-HVDC transferred the output power of the OWF. In addition, this study also checked on the DR-HVDC in transient state. Although the disconnection fault of the DC transmission line, which is one of the common incidents of an HVDC, occurred, the unidirectional HVDC in the Jeju Island power system was able to compensate for the dropped active power of the OWF by a fast response characteristic in the third case. This study also analyzed the ground fault impact of an offshore AC grid in the last case. Then, the analysis confirmed that the DR-HVDC can be helpful in reducing fault impact. From the results in steady and transient states, although the DR-HVDC was operated by using an uncontrolled rectifier at the offshore station and only one MMC at the onshore side, it was able to play a role as a conventional HVDC even in the small and isolated power system. Consequently, the application of the DR-HVDC in a small power system is reasonable to reduce costs and increase the reliability of an OWF instead of a conventional HVDC. If in the future, the new OWF in the Jeju Island power system is connected with a DR-HVDC, the power system will be more stable than the AC connection method and save costs than conventional HVDC topology.

Author Contributions: Conceptualization: S.-H.S. and E.-H.K. Methodology: S.H.C. and M.H.K. Resources: S.H.C., S.-H.S., and E.-H.K. Writing—original draft preparation: S.H.C. and M.H.K. Writing—review and editing: S.H.C. and E.-H.K. Supervision: S.-H.S. and E.-H.K.

Funding: This research was funded by KETEP and KEPCO, respectively.

Acknowledgments: This work was supported by the Korea Institute of Energy Technology Evaluation and Planning (KETEP) and the Ministry of Trade, Industry & Energy (MOTIE) of the Republic of Korea (No. 20173010024890) and Korea Electric Power Corporation (Grant number: R18XA03), respectively.

Conflicts of Interest: The authors declare no conflict of interest.

References

1. Jeju Special Self-Governing Province. *Carbon Free Island Jeju by 2030, Policy Report*; Jeju Special Self-Governing Province: Jeju Island, Korea, 2012.
2. Muyeen, S.M.; Takahashi, R.; Murata, T.; Tamura, J. Multi-Converter Operation of Variable Speed Wind Turbine Driving Permanent Magnet Synchronous Generator during Network Fault. In Proceedings of the 2008 International Conference on Electrical Machines and System, Tokyo, Japan, 15–18 November 2009. [CrossRef]
3. Bahrman, M.P. HVDC transmission overview. In Proceedings of the 2008 IEEE PES Transmission and Distribution Conference and Exposition, Chicago, USA, 21–24 April 2008. [CrossRef]
4. Reed, G.F.; Al Hassan, H.A.; Korytowski, M.J.; Lewis, P.T.; Grainger, B.M. Comparison of HVAC and HVDC solutions for offshore wind farms with a procedure for system economic evaluation. In Proceedings of the 2013 IEEE Energytech, Cleveland, OH, USA, 21–23 May 2013; pp. 1–7. [CrossRef]
5. Liu, H.; Chen, Z. Contribution of VSC-HVDC to frequency regulation of power system with offshore wind generation. *IEEE Trans. Energy Convers.* **2015**, *30*, 918–926. [CrossRef]
6. Elliott, D.; Bell, K.R.W.; Finney, S.J.; Adapa, R.; Brozio, C.; Yu, J.; Hussain, K. A comparison of AC and HVDC options for the connection of offshore wind generation in Great Britain. *IEEE Trans. Power Deliv.* **2016**, *31*, 798–809. [CrossRef]
7. Parastar, A.; Seok, J.-K. High-Power-Density Power Conversion System for HVDC-Connected Offshore Wind Farms. *J. Power Electron.* **2013**, *13*, 737–745. [CrossRef]
8. Cristian, N.; Hans-Gunter, E.; Sven, A.; Friedemann, A. Auxiliary Power Supply in a FixRef Controlled Offshore Wind Power Plant with Diode Rectifier HVDC Transmission. In Proceedings of the 16th Int'l Wind Integration Workshop, Berlin, Germany, 25–27 October 2017.
9. Seman, S.; Zurowski, R.; Taratoris, C. Interconnection of Advanced Type 4 WTGs with Diode rectifier based HVDC Solution and Weak AC Grids. In Proceedings of the 14th Wind Integration Workshop, Brussels, Belgium, 20–22 October 2015.
10. Prignitz, C.; Eckel, H.G.; Rafoth, A. FixRef Sinusoidal Control in Line Side Converters for Offshore Wind Power Generation. In Proceedings of the 2015 IEEE 6th International Symposium on Power Electronics for Distributed Generation System, Aachen, Germany, 22–25 June 2015. [CrossRef]

11. Marquardt, R.; Lesnicar, A.; Hildinger, J. Modulares stromrichterkonzept für netzkupplungsanwendungen bei hohen spannungen. In *Proceedings of the ETG Conference*; ETG-Fachtagung: Bad Nauheim, Germany, 2002.
12. Lesnicar, A.; Marquardt, R. An innovative modular multilevel converter topology suitable for a wide power range. In Proceedings of the Power Tech Conference, Bologna, Italy, 23–26 June 2003. [CrossRef]
13. Ilves, K.; Antonopoulos, A.; Norrga, S.; Nee, H.-P. Steady-state analysis of interaction between harmonic components of arm and line quantities of modular multilevel converters. *IEEE Trans. Power Electron.* **2012**, *27*, 57–68. [CrossRef]
14. Song, Q.; Liu, W.; Li, X.; Rao, H.; Xu, S.; Li, L. A steady-state analysis method for a modular multilevel converter. *IEEE Trans. Power Electron.* **2013**, *28*, 3702–3713. [CrossRef]
15. Zhang, Y.; Adam, G.P.; Lim, T.C.; Finney, S.J.; Williams, B.W. Analysis and Experiment Validation of a Threelevel Modular Multilevel Converters. In Proceedings of the 8th International Conference on Power Electronics—ECCE Asia, Jeju, Korean, 29 May–2 June 2011. [CrossRef]
16. Tu, Q.; Xu, Z.; Chang, Y.; Guan, L. Suppressing DC voltage ripples of MMC-HVDC under unbalanced grid conditions. *IEEE Trans. Power Deliv.* **2012**, *27*, 1332–1338. [CrossRef]
17. Guam, M.; Xu, Z. Modeling and control of a modular multilevel converter-based HVDC system under unbalanced grid conditions. *IEEE Trans. Power Electron.* **2012**, *27*, 4858–4867. [CrossRef]
18. Rohner, S.; Bernet, S.; Hiller, M.; Sommer, R. Modulation, losses, and semiconductor requirements of modular multilevel converters. *IEEE Trans. Ind. Electron.* **2010**, *57*, 2633–2642. [CrossRef]
19. Guan, M.; Xu, Z.; Chen, H. Control and modulation strategies for modular multilevel converter based HVDC system. In Proceedings of the IECON 2011—37th Annual Conference on IEEE Industrial Electronics Society, Melbourne, VIC, Australia, 7–10 November 2011. [CrossRef]
20. Yang, X.; Li, J.; Wang, X.; Fan, W.; Zheng, T.Q. Circulating current model of modular multilevel converter. In Proceedings of the 2011 Asia-Pacific Power and Energy Engineering Conference (APPEEC), Wuhan, China, 25–28 March 2011.

Article

Radiation View Factor for Building Applications: Comparison of Computation Environments

Marzia Alam [1,*], Mehreen Saleem Gul [1] and Tariq Muneer [2]

1 School of Energy, Geoscience, Infrastructure and Society, Heriot-Watt University, Boundary Rd N, Edinburgh EH14 4AS, UK; m.gul@hw.ac.uk
2 School of Engineering and The Built Environment, Edinburgh Napier University, 10 Colinton Rd, Edinburgh EH10 5DT, UK; T.Muneer@napier.ac.uk
* Correspondence: ma410@hw.ac.uk; Tel.: +44-07544930047

Received: 24 September 2019; Accepted: 8 October 2019; Published: 10 October 2019

Abstract: Computation of view factors is required in several building engineering applications where radiative exchange takes place between surfaces such as ground and vertical walls or ground and sloping thermal or photovoltaics collectors. In this paper, view factor computations are performed for bifacial solar photovoltaic (PV) collectors based on the finite element method (FEM) using two programming languages known as Microsoft Excel-Visual Basic for Applications (VBA) and Python. The aim is to determine the computer response time as well as the performance of the two languages in terms of accuracy and convergence of the numerical solution. To run the simulations in Python, an open source just-in-time (JIT) compiler called Numba was used and the same program was also run as a macro in VBA. It was observed that the simulation response time significantly decreased in Python when compared to VBA. This decrease in time was due to the increase in the total number of iterations from 400 million to 250 billion for a given case. Results demonstrated that Python was 71–180 times faster than VBA and, therefore, offers a better programming platform for the view factor analysis and modelling of bifacial solar PV where computation time is a significant modelling challenge.

Keywords: building energy exchange; view factor; Python programming language; bifacial solar photovoltaic (PV)

---

1. Introduction

Buildings consume a great deal of energy to maintain comfortable conditions within their enclosures. The design of heating and cooling systems for buildings require a detailed assessment of the temperature of the constituting envelope. Thus, the temperature of the walls, roofs, floors, and glazing surfaces are required. Buildings within urban areas inevitably have radiative exchanges with other buildings, sky, and ground. Radiation heat transfer has many applications within the building services sector. Some of the building engineering applications of radiative exchange between surfaces are:

- Ground and vertical walls or windows (Figure 1).
- A light-pipe illuminating the floor or desk spaces (Figure 2).
- Ground and sloping thermal or photovoltaic (PV) collectors.
- Canyon space in densely populated urban areas with radiative exchange between parallel walls and/or windows.
- Radiative exchange between orthogonal walls and/or windows.
- Wet heating system's radiator and indoor surface of the walls.

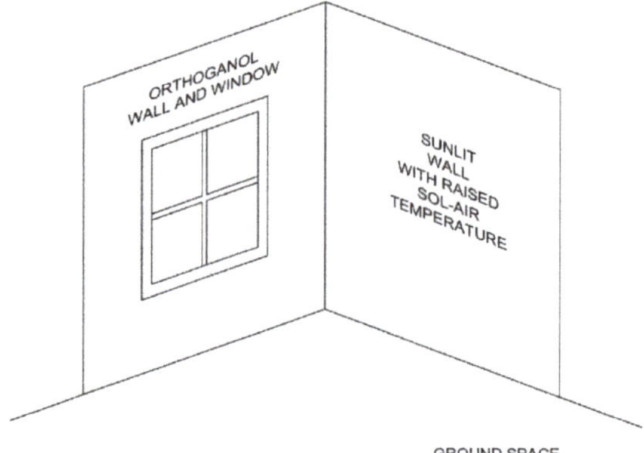

Figure 1. Ground and vertical walls or windows.

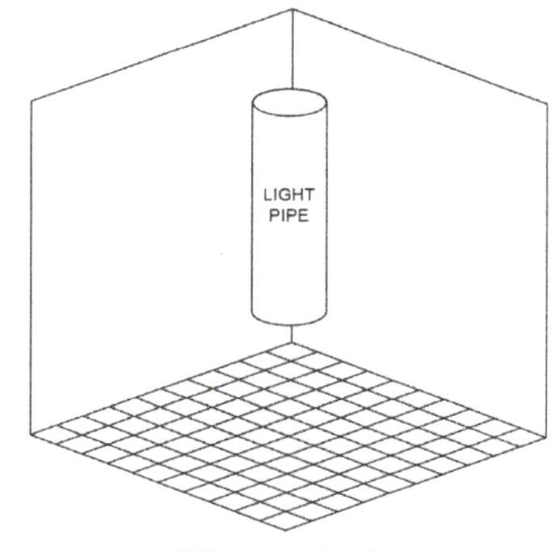

Figure 2. A light-pipe illuminating floor or desk spaces.

In all the above applications' computation of the radiation view factor is required. There have been significant research studies related to energy transfer in building applications such as quantification of available façade areas for installation of building integrated solar PV [1], determination of solar heat gain for daylighting studies [2], theoretical modelling of the view factor to determine the diffused components of solar irradiance [3], and predictive simulation tool to determine the radiance value of a façade in the street canyon [4].

In this article, we have considered the radiative energy exchange of solar photovoltaic collectors such as bifacial solar photovoltaics (PV), which is gaining popularity both for the field level and rooftop building applications [5]. It is a promising technology that increases the production of electricity per square meter of the PV module using light absorption from the albedo [6]. Bifacial solar cells simultaneously collect photons from incident and albedo radiation reaching both the front side and the back side of a solar module whereas traditional monofacial solar cells only collect photons reaching the front side of the device. Cuevas et al. [7] showed that an increase of 50% in electric power generation can be obtained by simultaneously collecting direct and albedo radiation from the rooftop and surroundings around a module. Consequently, it was established that bifacial solar cells can increase the power density of PV modules compared to mono-facial cells while reducing area-related costs for PV systems [8]. Currently, there are two main challenges this technology is facing, which are

(1) a lack of an adequate simulation tool that can help understand how the rear side of the PV interact with the reflected sunlight and (2) how to make the technology bankable. The success of the technology will depend on how well the bifacial PV is understood to the research community. Therefore, it would be easier to make the technology commercially profitable. Hence, the need for developing a simulation tool to understand the energy yield of bifacial solar PV is inevitable. There are different simulation approaches to determine the ground reflected radiation that is incident upon tilted bifacial solar photovoltaics (PV) such as the view factor model, the ray tracing model, and empirical modelling [9]. There has been ongoing research by various scientific communities. For instance, National Renewable Energy Laboratory (NREL) in collaboration with Sandia National Laboratories have developed and compared the view factor model and the ray tracing model to evaluate the back-surface irradiance [10]. However, computation time remain an issue, which is still a modelling challenge that requires further investigation. Conceptually, the back-surface irradiance is composed of direct irradiance, structure reflected irradiance, sky diffused irradiance, and ground reflected irradiance. In this article, we have followed the view factor modelling-based approach to evaluate the radiative energy transfer between the ground (reflecting surface) and the bifacial solar PV (collecting surface).

The present research team had developed a computer code for view factor analysis using a finite-element grid, which can handle an irregular horizon [11]. The software was developed based on a numerical solution of the view factor integral within the Microsoft Excel-Visual Basic for Applications (VBA) environment considering mono-facial solar PV collectors as an example. However, in this work, an improved, simplified brute force algorithm is adopted to determine the view factor for uniform surfaces by utilizing the finite element method (FEM). FEM is one of the most powerful tools for analyzing energy exchange between surfaces. It can be applied for view factor (VF) evaluation. This method is very robust and, yet, it may require excessive computer time. For finite element analysis, the object-oriented computing environment is already proven to be efficient in various areas such as Structural Engineering [12], Chemical Engineering, and Mathematics. In this article, we have implemented the view factor (VF) code in VBA and the Python environment and have compared the computation speed of the simulations in both platforms.

Python is chosen because it has already been accepted by the research community for its ease of use and an open access user-friendly platform. Availability of the open source library has made it eligible for numerous scientific applications such as power systems [13], energy analysis [14] mathematics, and fluid dynamics [15]. VBA has also been deployed as a user-friendly tool in various applications such as in heat transfer, in solar PV applications [16,17], and in the field of agriculture. For example, a milk-producing dairy model was implemented in VBA to control the operation of the dairy farm, which has a significant benefit to the farmers and researchers in the field. However, it takes seven days to run the simulation [18]. Therefore, despite its effectiveness as a design tool, computation time has always been an issue for VBA [19] whereas Python, which is a high-level, benchmark programming language, can outperform VBA in terms of computation speed.

In this paper, we have compared the potentials of the simulation platform to develop the code further for bifacial solar PV collectors. Two simulation platforms are used to quantify the amount of ground reflected radiation received by the solar collectors. We focused on the modelling of computationally intensive view factor analysis between reflecting (ground) and collecting (bifacial solar PV) surface and determined the duration of simulation response time. The paper is organized as follows. The concept of view factor modelling is briefly discussed in Section 2. Section 3 provides the overview of the simulation model. The computation performance of two simulation platforms, Python and Microsoft Excel-Visual Basic for Applications (VBA), is demonstrated in Section 4. We have presented an application related to generation of solar PV electricity and its enhancement by using reflective films placed near the horizon of PV modules. The object is to obtain a numerical procedure for the radiative exchange between the foreground and PV modules. The technique can be applied to a bifacial solar PV cell, which is an emerging technology. The novelty of this work is that we have provided quantitative data that demonstrates faster convergence and accuracy offered by the Python software environment in the context of modelling bifacial solar PV's energy yield where computation time is a significant modelling challenge. Lastly, the paper ends with concluding remarks in Section 5.

2. View Factor Modelling Concept

In our work, we simulated the mathematical model of the radiation view factor as a discrete model in the computing platform by applying finite element method. To understand the concept of view factor modelling, let us consider two rectangular surfaces A_1 and A_2 with surface dimensions $a \times b$ and $c \times b$ and Φ is the tilt angle between the surfaces (Figure 3a) and $\beta = \pi - \Phi$. In radiative heat transfer, the view factor can be defined as the proportion of irradiance, which leaves the emitting surface A_2 and strikes the receiving/collector surfaces A_1 denoted by F_{A2-A1}. From the mathematical equation of the view factor [11], we may write:

$$F_{A_2-A_1} = \frac{1}{bc}\int_{x_1}^{c}\int_{y_1}^{b}\int_{x_2}^{a}\int_{y_2}^{b} \frac{x_1 x_2\, sin^2\beta}{\pi\left[x_1{}^2 + x_2{}^2 + 2x_1x_2\, cos\beta + (y_1 - y_2)^2\right]^2}\, dy_2 dx_2 dy_1 dx_1. \quad (1)$$

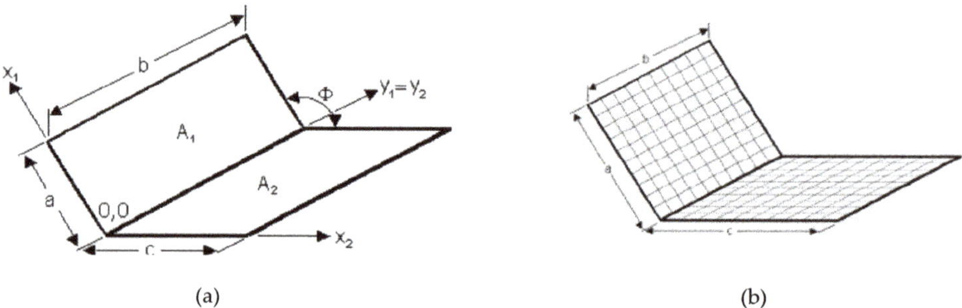

Figure 3. Radiation view factor analysis: (**a**) Two rectangles with one common edge and (**b**) the reflecting and receiving surface with uniform grids [11].

In the present work, Equation (1) has been used to compute the view factor for different geometries of the receiving (solar collectors) and reflecting (ground) surface, which share a common edge. Both the emitting and receiving surfaces are considered as uniform grids where all the cells within the surface are of the same dimensions and aspect ratios (Figure 3b). The tilt angle between the two surfaces varied from 30°–150°. The height of the receiving and emitting surface is denoted by 'a' and 'c', respectively, and the common edge length is denoted by 'b'.

One of the approaches to examine the solution of the discrete model is to observe the convergence of the computed solution toward the analytical solution (if it exists) of the mathematical model. This approach is known as verification. For our paper, the analytical solution provided by Feingold [20] is considered as the benchmark solution for the purpose of verification. However, any physical model when interpreted by a discrete model, the solution error, or computation error is an important phenomenon that needs to be properly understood. For a uniform grid, the difference between the analytical solution and the computed solution represents the error. In this case, we have set up the finite element procedure to solve the problem repeatedly with uniform meshes designed to determine the view factor and the error.

There are different adaptive finite element techniques available to estimate the error such as local refinement or h-refinement, relocating or r-refinement, and locally varying the polynomial degree known as p-refinement. For the purpose of verification of our model, an h-adaptive refinement technique is applied where the h denotes the element size or resolution of the grid, which we used to control the error. Decreasing h leads to reduction of the error. As h approaches zero, the numerical solution converges toward its analytical value. The error can be derived from Equation (2) where error (E_r) is a function of element size h_i. VF_{ex} represents the analytical value of the view factor and VF_{hi} is the simulated value of the view factor at element size h_i. C is an unknown constant with the leading term $h_i{}^{\beta}$ and exponent β is the rate of convergence [21].

$$\lim_{h_{i=1,2,3,4..}\to 0} Er(h_i) = VF_{ex} - VF_{h_i} \approx Ch_i^{\beta} \; for \; \beta > 0 \tag{2}$$

If the condition $\frac{h_1}{h_2} = \frac{h_2}{h_3}$ holds, Equation (2) can be further derived as:

$$\frac{VF_{ex} - VF_{h1}}{VF_{ex} - VF_{h2}} = \frac{h_1^{\beta}}{h_2^{\beta}} \; and \; \frac{VF_{ex} - VF_{h2}}{VF_{ex} - VF_{h3}} = \frac{h_2^{\beta}}{h_3^{\beta}} \tag{3}$$

The element size determines the number of iterations the simulation must run. The parameters considered for view factor models are presented in Table 1 below.

Table 1. View factor modeling parameters.

Surface Dimensions (m)	Element Size h_i (m)	Φ (°)
a = 2 m, b = 1 m, c = 2 m	0.01	30
a = 1 m, b = 1 m, c = 1 m	0.008	60
a = 0.6 m, b = 1 m, c = 0.4 m	0.004	90
a = 0.4 m, b = 1 m, c = 0.6 m	0.002	120
a = 0.4 m, b = 0.4 m, c = 0.4 m	—	150

3. Overview of the Simulation Model

The numerical view factor model has been executed both in Python and VBA environments. The simulation utilizes one of the most efficient Python libraries for numerically intensive computing named Numba, which can be loaded by a program as the CPython interpreter. The code is written based on Numba just-in-time (JIT) compiler, which creates a specialized loop in the machine code. The just-in-time or @ JIT is a decorator that is utilized as a function. When this function is called, the decorator analyzes the argument and creates a specialized version of the function [22]. This code is run on the 'nopython' mode, which compiles the code without accessing the Python C-API (application program interface). Python version 3.7 with Spyder integrated development environment (IDE) is used for the simulation. On the other hand, VBA adopts the run time library of the Visual Basic, which is compiled in a Microsoft packet code and the MS-Excel works as a host application that saves the code in a separate file such as .xlsx or .xlsm. A machine hosted by Excel run the intermediate code and the data is saved in a text file in XML format, which is readable by the user [23]. The simulations were run on Intel®Core ™ i7-7500U CPU @ 2.7 GHz-2.9 GHz Laptop.

Python and VBA both used a Microsoft Excel worksheet as an input and output file. The user can enter the input parameters, run the simulation, and calculate the simulation response time as well as observe the output, which is saved in an Excel file. Different simulation parameters can be set such as the length and height of the surfaces. For example, referred to Figure 3b, the angle between two surfaces sharing one common edge can be selected and varied for the same element size, h_i. The simulations were run at different resolutions of the element size defined as the iteration step in the program, which is varied as: h_1 = 0.01 m, h_2 = 0.008 m, h_3 = 0.004 m, and h_4 = 0.002 m. This iteration step defines the number of iterations the simulation needs to run, which can be calculated from Equation (4).

$$Iteration\ size = \frac{c}{h_i} \times \frac{b}{h_i} \times \frac{a}{h_i} \times \frac{b}{h_i} \tag{4}$$

Thus, iteration ranges of the simulations vary from 4.00×10^8 to 2.50×10^{11}. In this study, a brute force problem-solving approach is applied within a mathematical equation of the view factor (given by Equation (1) in Section 3), which aims to check all the possible interactions between the surfaces using four nested loops bounded by the surface dimensions [24]. The program flow chart is presented in Figure 4.

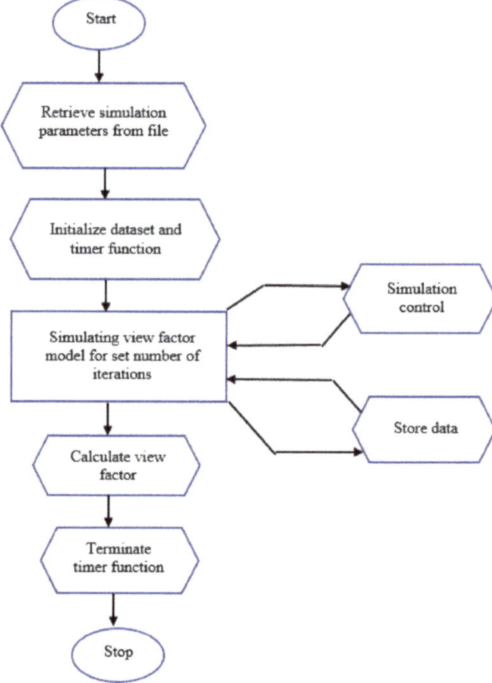

Figure 4. Flow chart of the simulation model.

4. Results and Discussion

The view factor simulation outputs are examined from different approaches, as explained below.

- The simulated view factor output is tested at different computation angles between collectors and reflective surfaces to verify the numerical model with the existing analytical solution.
- The reduction in the percentage (%) error of the view factor output is evaluated with an increasing number of iterations. An accuracy versus computation time dependency is shown in Section 4.2.
- Followed by Section 4.2, the convergence rate of the simulated output to the actual output is examined to determine the type of convergence (linear/quadratic/cubic).
- Next, the computation time versus the number of iterations are tested for specific surface dimensions to compare the improvement of the simulation response time in Python over VBA.
- Lastly, computation time variations with different surface dimensions are determined. Detailed results and analysis are discussed in the following sections. All simulations parameters and outputs are referred to Appendix A.

4.1. View Factor at Different Computation Angles Between Two Surfaces

This section attempts to verify the proposed view factor model at different tilt angles between the surfaces. The simulated results show a minor deviation from the existing analytical solutions. It has been observed that view factors are considerably influenced by the angle between the surfaces, as illustrated in Figure 5. Keeping the dimensions and iteration size constant, as the angle between two surface increases, a significant decrease in the view factor from a 0.51 maximum value (for this study) to the minimum value of 0.015 is observed. Parameters for this analysis are presented in Table 2 below.

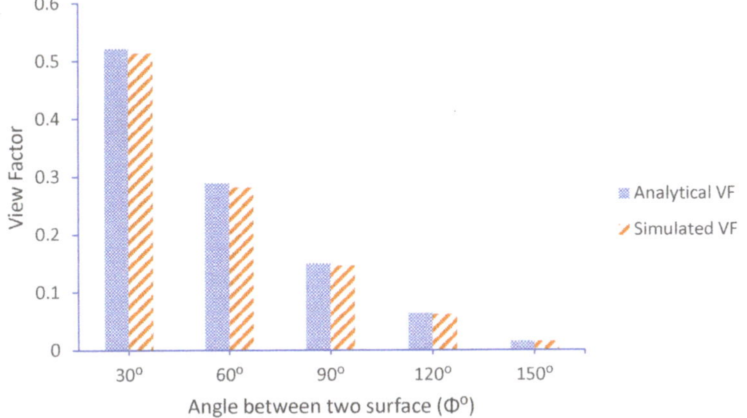

Figure 5. View factor at different angles between surfaces for a = 2 m, b = 1 m, and c = 1 m, and h_i = 0.01 m.

Table 2. View factor simulation parameters for element size h_i = 0.01 m and number of iterations = 4.0×10^8.

a (m)	b (m)	c (m)	Element Size h_i (m)	Φ (°)	Analytical VF	Simulated VF	Error (%)
2	1	2	0.01	30	0.521308	0.513849	1.4
2	1	2	0.01	60	0.288274	0.281650	2.3
2	1	2	0.01	90	0.149300	0.145292	2.7
2	1	2	0.01	120	0.063248	0.061372	3.0
2	1	2	0.01	150	0.015415	0.014937	3.1

Table 2 illustrates that, for element size h_i = 0.01 m, the percentage error varies from a minimum 1.4% to a maximum 3.1% depending on the value of angle Φ. The iteration size for this simulation is set to 4.0×10^8. The next section presents, how this error can be further reduced by decreasing the element size h_i, which, in turn, increases the number of iterations.

4.2. Accuracy Versus the Number of Iterations

The accuracy of the view factor output significantly increased with the number of iterations the simulation runs, which can be seen in Figure 6. Both Python and VBA achieved the same level of accuracy. For surface dimensions: a = 2 m, b = 1 m, c = 2 m, and Φ = 120°, the computation accuracy is 97% with 4.0×10^8 iterations. The numerical results approached the analytical solution at a maximum accuracy of 99.4% for 2.5×10^{11} iterations. To gain this, the iteration size is to be increased by a factor of 625. The simulation parameters for this study are outlined in Table 3 below.

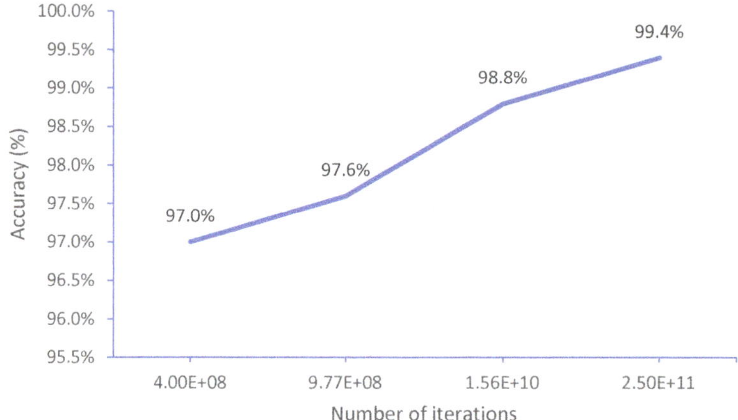

Figure 6. Accuracy versus the number of iterations for $a = 2$ m, $b = 1$ m and $c = 2$ m and $\Phi = 120°$.

Table 3. View factor simulation parameters at various element sizes.

a (m)	b (m)	c (m)	**Φ** (°)	Element Size, h_i (m)	Analytical VF	Simulated VF	Error (%)
2	1	2	120	0.01	0.063248	0.061372	3.0
2	1	2	120	0.008	0.063248	0.061743	2.4
2	1	2	120	0.004	0.063248	0.062491	1.2
2	1	2	120	0.002	0.063248	0.062868	0.59

4.3. Convergence to an Analytical Solution

The convergence performance of the view factor simulation to the analytical solutions are studied by comparing the output between the highest: $a = 2$ m, $b = 1$ m, $c = 2$ m, $\Phi = 120°$ and the lowest: $a = 0.4$ m, $b = 1$ m and $c = 0.4$ m, $\Phi = 120°$ surface dimensions. For the same number of iterations, maximum accuracy achieved for Figure 7a was up to 99.4% whereas, for Figure 7b, the accuracy was 98.79%. Comparing both simulation responses, it can be inferred that, to reach the same level of convergence, the element size needs to be further reduced for surface dimensions: $a = 0.4$ m, $b = 1$ m, and $c = 0.4$ m; $\Phi = 120°$. In addition, the convergence type of the model is found to be linear with the β value of approximately 1. The rate of convergence is derived from Equation (3) considering the element size of 0.004 m and 0.002 m. Table 4 summarizes the results.

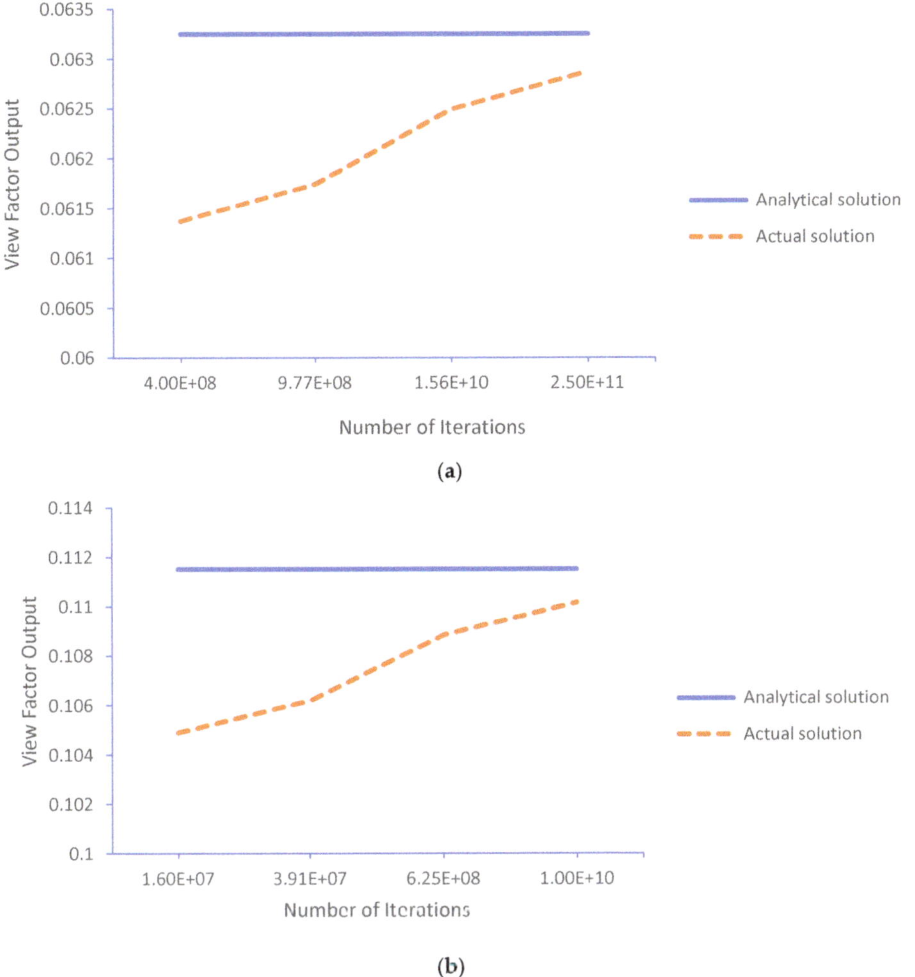

(b)

Figure 7. (**a**). Convergence to an analytical solution for $a = 2$ m, $b = 1$ m, $c = 2$ m, and $\Phi = 120°$. (**b**). Convergence to an analytical solution for $a = 0.4$ m, $b = 1$ m, $c = 0.4$ m, and $\Phi = 120°$.

Table 4. Convergence of the view factor output.

Surface Dimension (m)	Analytical VF	Simulated VF $h_i = 0.004$ m	Simulated VF $h_i = 0.002$ m	Convergence Rate (β)
a = 0.4, b = 1, c=0.4	0.111512	0.108828	0.110164	0.993564
a = 0.4, b = 1, c = 0.6	0.085512	0.083668	0.084586	0.993755
a = 0.6, b = 1, c = 0.4	0.128269	0.125502	0.126879	0.993238
a = 1, b = 1, c = 1	0.086615	0.085348	0.085979	0.994318
a = 2, b = 1, c = 2	0.063248	0.062491	0.062868	0.994294

4.4. Computation Time versus Number of Iterations

The preceding sections have shown, although both Python and VBA gained the same accuracy with an increasing number of iterations, VBA simulation took more time than Python to reach the same level of accuracy. Figure 8 presents the computation time in seconds for both programs. We see that Python took 3.52 s to run 4.0×10^8 iterations whereas VBA required 454 s. For all other iterations, Python outperformed VBA in terms of computation speed, and it was about 129–180 times faster than VBA. Simulation parameters for these computations can be found in Table 5.

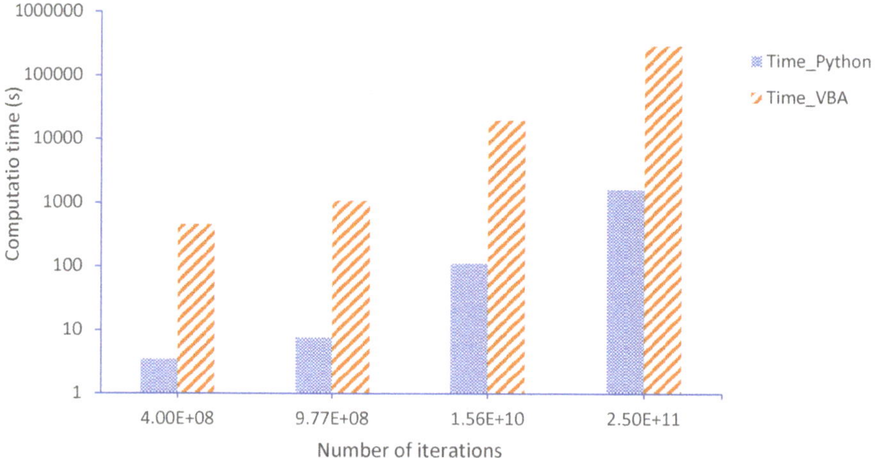

Figure 8. Computation time versus the number of iterations for a = 2 m, b = 1 m, and c = 1 m and Φ = 120°.

Table 5. Simulation parameters for computation time at different element sizes.

a (m)	b (m)	c (m)	Φ (°)	Element Size h_i (m)
2	1	2	120	0.01
2	1	2	120	0.008
2	1	2	120	0.004
2	1	2	120	0.002

4.5. Computation Time for Different Surface Dimensions

The view factor simulation response time increased with the increasing length and height of surface dimensions. Table 6 summarizes results of computation time variations among different surface dimensions. The results show, in VBA, the lowest computation time is 24 s for the surface dimensions: a = 0.4 m, b = 1 m, and c = 0.4 m. This computation time increased by a factor of 19 for the maximum dimension considered in the simulation: a = 2 m, b = 1 m, and c = 2 m. Nevertheless, Python appears to be 71–129 times faster than VBA at a varying surface length and height.

Table 6. Computation time at different surface lengths and heights for Φ = 120° and h_i = 0.01 m.

Surface Dimension (m)	Element Size (m)	Analytical VF	Simulated VF	Time_Python (s)	Time_VBA (s)
a = 0.4, b = 1, c = 0.4	0.01	0.111512	0.104891	0.34	24
a = 0.4, b = 1, c = 0.6	0.01	0.085512	0.080957	0.41	43
a = 0.6, b = 1, c = 0.4	0.01	0.128269	0.121436	0.45	43
a = 1, b = 1, c = 1	0.01	0.087615	0.083481	0.92	222
a = 2, b = 1, c = 2	0.01	0.063248	0.061372	3.52	454

A finite element computation of the view factor is a time critical application. The present view factor model offers a quantitative computational advantage, i.e., grid surface non-uniformity can be handled quite easily. Other advantages offered by this work are: (a) the view factor code is much faster for geometries encountered in most solar energy and building energy exchange applications, and (b) the view factor approach allows for much shorter computation time, particularly if handled in the Python environment. Note that implementation of other alternative method for instance, ray tracing as a modelling tool is more complex compared to the view factor concept.

The computing performance not only depends on the type of computer used but also on the simulation algorithm and its implementation platform, i.e., programming language. Though short response time is one of the benchmark criteria of the performance efficiency, there are additional performance matrixes, which can be achieved with fast computation speed. These include high throughput, minimum utilization of resources [25], higher reliability [26], low power consumption, scalability, and opportunity of performance tuning. For our study, one downside of VBA was that it took about five consecutive days to run 250 billion iterations, which can have substantial damage on computer health. Moreover, utilization of computer resources has been very high, which required lots of power consumption. In addition to this, the scope of performance tuning in VBA is also limited, which reduces the scalability of the number of iterations the simulation can run. In all these aspects, Python can outperform VBA, which make it a suitable option for present day scientific and numeric computing. The existing work demonstrates the benefits of using Python for view factor analysis for solar PV applications, which is an improvement over our previous work where we used VBA for such an analysis.

5. Conclusions

In this paper, we have computed the value of the radiation view factor to determine the reflected solar irradiance reaching the rear side of the bifacial solar PV. We have verified the results with the existing analytical solutions. In this scenario, we focus on the computing performance to examine the improvement in computation speed of Python as compared to VBA. It has been shown that, Python can be used more effectively than VBA for radiation view factor analysis between two surfaces. With the utilization of an appropriate mathematical library, computation time was significantly reduced by 71–180 times for Python when compared with VBA. This improvement in computation speed not only saves time, but also provides an optimized design tool for the research. An important finding of the view factor simulation is that, as the element size of the finite element grid decreased, the computed output converged to the analytical view factor value. Thus, the simulation accuracy could be achieved up to 99.4% for the maximum number of iterations considered for this paper, i.e., 250 billion and the response time of the simulation in Python and VBA was 1628.51 s and 292,714 s, respectively. The application presently considered in this article are for relatively small areas of the reflecting and the receiving surfaces. In an actual industrial environment where the designer will deal with multi-gigawatt solar PV farms, that may employ enhanced reflections near the horizon. In that case, to model such large-scale systems, the number of iterations the computer simulations have to run will increase to the order of a few quadrillion or more. Therefore, the importance of faster code written in a computation environment such as Python will be of great benefit to the PV system designers. Hence, it is concluded that, Python can be utilized as a reliable simulation tool to develop the code further for bifacial solar PV research.

Author Contributions: Conceptualization: T.M., M.S.G., and M.A. Methodology: M.A., T.M., and M.S.G. Formal analysis: M.S.G., T.M., and M.A. Coding: T.M., M.A. Resources: T.M., M.S.G., and M.A. Data curation: M.A., T.M., and M.S.G. Writing—original draft preparation: M.A., M.S.G., and T.M. Writing—review and editing: T.M., M.S.G., and M.A. Supervision: M.S.G. and T.M.

Funding: This project is being run in collaboration with four partners: Energy Technology Partnership (ETP-Project 161) Scotland, Wood Group-Clean Energy, Heriot-Watt University and Edinburgh Napier University, UK.

Conflicts of Interest: The authors declare no conflict of interest.

Appendix A View Factor Calculation Tables at Different Element Sizes

Table A1. View factor simulation data for element size $h_i = 0.01$ m.

a (m)	b (m)	c (m)	Φ (°)	Iteration	Analytical VF	Simulated VF	Error (%)	Time_VBA (s)	Time_Python (s)
2	1	2	30	4.0×10^8	0.521308	0.513849	1.4	415	2.77
2	1	2	60	4.0×10^8	0.288274	0.281650	2.3	410	2.75
2	1	2	90	4.0×10^8	0.149300	0.145292	2.7	408	2.87
2	1	2	120	4.0×10^8	0.063248	0.061372	3.0	454	2.74
2	1	2	150	4.0×10^8	0.015415	0.014937	3.1	410	2.80
1	1	1	90	1.0×10^8	0.200044	0.193529	3.3	143	0.87
0.4	1	0.4	120	1.6×10^7	0.111512	0.104891	5.9	24	0.34
1	1	1	120	1.0×10^8	0.087615	0.083481	3.6	222	0.92
0.6	1	0.4	120	2.4×10^7	0.128269	0.121436	5.3	43	0.45
0.4	1	0.6	120	2.4×10^7	0.085512	0.080957	5.3	43	0.41
0.4	1	0.6	30	2.4×10^7	0.518407	0.508234	2.0	44	0.38
0.6	1	0.4	30	2.4×10^7	0.777610	0.762351	2.0	35	0.39

Table A2. View factor simulation data for element size $h_i = 0.008$ m.

a (m)	b (m)	c (m)	Φ (°)	Iteration	Analytical VF	Simulated VF	Error (%)	Time_VBA (s)	Time_Python (s)
2	1	2	30	9.76×10^8	0.521308	0.515332	1.1	779	7.98
2	1	2	60	9.76×10^8	0.288274	0.282965	1.8	780	7.35
2	1	2	90	9.76×10^8	0.149300	0.146086	2.2	891	7.67
2	1	2	120	9.76×10^8	0.063248	0.061743	2.4	773	7.35
2	1	2	150	9.76×10^8	0.015415	0.015026	2.5	790	7.40
1	1	1	90	2.44×10^8	0.200044	0.194818	2.6	428	1.98
0.4	1	0.4	120	3.96×10^7	0.111512	0.106192	4.8	55	0.48
1	1	1	120	2.44×10^8	0.087615	0.084099	2.9	429	1.95
0.6	1	0.4	120	5.86×10^7	0.128269	0.122781	4.3	90	0.61
0.4	1	0.6	120	5.86×10^7	0.085512	0.081854	4.3	91	0.61
0.4	1	0.6	30	5.86×10^7	0.518407	0.510242	1.6	92	0.60
0.6	1	0.4	30	5.86×10^7	0.777610	0.765363	1.6	77	1.17

Table A3. View factor for element size $h_i = 0.004$ m.

a (m)	b (m)	c (m)	Φ (°)	Iteration	Analytical VF	Simulated VF	Error (%)	Time_VBA (s)	Time_Python (s)
2	1	2	30	1.56×10^{10}	0.521308	0.518310	0.58	18971	111.14
2	1	2	60	1.56×10^{10}	0.288274	0.285609	0.92	17010	110.85
2	1	2	90	1.56×10^{10}	0.149300	0.147685	1.08	14997	110.8
2	1	2	120	1.56×10^{10}	0.063248	0.062491	1.20	19407	110.78
2	1	2	150	1.56×10^{10}	0.015415	0.015219	1.27	17493	110.79
1	1	1	90	3.90×10^9	0.200044	0.197415	1.31	7180	27.74
0.4	1	0.4	120	6.25×10^8	0.111512	0.108828	2.41	1120	4.55
1	1	1	120	3.90×10^9	0.087615	0.085348	1.46	4649	27.74
0.6	1	0.4	120	9.37×10^8	0.128269	0.125502	2.16	737	6.66
0.4	1	0.6	120	9.37×10^8	0.085512	0.083668	2.16	738	6.71
0.4	1	0.6	30	9.37×10^8	0.518407	0.514296	0.79	751	6.65
0.6	1	0.4	30	9.37×10^8	0.777610	0.771444	0.79	818	6.71

Table A4. View factor for element size h_i = 0.002 m.

a (m)	b (m)	c (m)	Φ (°)	Iteration	Analytical VF	Simulated VF	Error (%)	Time_VBA (s)	Time_Python (s)
2	1	2	30	2.50×10^{11}	0.521308	0.519806	0.28	322099	1732.83
2	1	2	60	2.50×10^{11}	0.288274	0.286939	0.46	257484	1726.24
2	1	2	90	2.50×10^{11}	0.149300	0.148490	0.54	270788	1632.94
2	1	2	120	2.50×10^{11}	0.063248	0.062868	0.59	292714	1628.51
2	1	2	150	2.50×10^{11}	0.015415	0.015317	0.63	309316	1627.28
1	1	1	90	6.25×10^{10}	0.200044	0.198725	0.65	74904	405.54
0.4	1	0.4	120	1.10×10^{10}	0.111512	0.110164	1.20	23990	65.5
1	1	1	120	6.25×10^{10}	0.087615	0.085979	0.73	78338	407.8
0.6	1	0.4	120	1.50×10^{10}	0.128269	0.126879	1.08	73302	103.16
0.4	1	0.6	120	1.50×10^{10}	0.085512	0.084586	1.082	70983	102.52
0.4	1	0.6	30	1.50×10^{10}	0.518407	0.516343	0.39	45392	103.33
0.6	1	0.4	30	1.50×10^{10}	0.777610	0.774515	0.39	31975	103.42

References

1. Costanzo, V.; Yao, R.; Essah, E.; Shao, L.; Shahrestani, M.; Oliveira, A.C.; Araz, M.; Hepbasli, A.; Biyik, E. A Method of Strategic Evaluation of Energy Performance of Building Integrated Photovoltaic in the Urban Context. *J. Clean. Prod.* **2018**, *184*, 82–91. [CrossRef]
2. Zheng, C.; Wu, P.; Costanzo, V.; Wang, Y.; Yang, X. Establishment and Verification of Solar Radiation Calculation Model of Glass Daylighting Roof in Hot Summer and Warm Winter Zone in China. *Procedia Eng.* **2017**, *205*, 2903–2909. [CrossRef]
3. Maor, T.; Appelbaum, J. View Factors of Photovoltaic Collector Systems. *Sol. Energy* **2012**, *86*, 1701–1708. [CrossRef]
4. Rubio-Bellido, C.; Pulido-Arcas, J.A.; Sánchez-Montañés, B. A Simplified Simulation Model for Predicting Radiative Transfer in Long Street Canyons under High Solar Radiation Conditions. *Energies* **2015**, *8*, 13540–13558. [CrossRef]
5. Cha, H.L.; Bhang, B.G.; Park, S.Y.; Choi, J.H.; Ahn, H.K. Power Prediction of Bifacial Si PV Module with Different Reflection Conditions on Rooftop. *Appl. Sci.* **2018**, *8*, 1752. [CrossRef]
6. Guerrero-Lemus, R.; Vega, R.; Kim, T.; Kimm, A.; Shephard, L.E. Bifacial Solar Photovoltaics—A Technology Review. *Renew. Sustain. Energy Rev.* **2016**, *60*, 1533–1549. [CrossRef]
7. Cuevas, A.; Luque, A.; Eguren, J.; del Alamo, J. 50 Per Cent More Output Power from an Albedo-Collecting Flat Panel Using Bifacial Solar Cells. *Sol. Energy* **1982**, *29*, 419–420. [CrossRef]
8. Kreinin, L.; Bordin, N.; Karsenty, A.; Drori, A.; Eisenberg, N. Experimental Analysis of the Increases in Energy Generation of Bifacial over Mono-Facial PV Modules. In Proceedings of the 26th Europe Photovoltaic Solar Energy Conference Exhibition, Hambourg, Genmany, 5–9 September 2011; pp. 3140–3143.
9. Shoukry, I.; Berrian, D.; Libal, J.; Haffner, F. Simulation Models for Energy Yield Prediction of Bifacial Systems. In *Bifacial Photovoltaics: Technology, Applications and Economics*; Institution of Engineering and Technology (IET): Stevenage, UK, 2018. [CrossRef]
10. Hansen, C.W.; Stein, J.S.; Deline, C.; Macalpine, S.; Marion, B.; Asgharzadeh, A.; Toor, F. Analysis of Irradiance Models for Bifacial PV Modules. In Proceedings of the 2017 IEEE 44th Photovoltaic Specialist Conference, PVSC, Washington, DC, USA, 25–30 June 2017. [CrossRef]
11. Muneer, T.; Ivanova, S.; Kotak, Y.; Gul, M. Finite-Element View-Factor Computations for Radiant Energy Exchanges. *J. Renew. Sustain. Energy* **2015**, *7*, 033108. [CrossRef]
12. Rypl, D.; Patzák, B. From the Finite Element Analysis to the Isogeometric Analysis in an Object-Oriented Computing Environment. *Adv. Eng. Softw.* **2012**, *44*, 116–125. [CrossRef]
13. Milano, F. A Python-Based Software Tool for Power System Analysis. In Proceedings of the IEEE Power and Energy Society General Meeting, Vancouver, BC, Canada, 21–25 July 2013. [CrossRef]
14. Zoder, M.; Balke, J.; Hofmann, M.; Tsatsaronis, G. Simulation and Exergy Analysis of Energy Conversion Processes Using a Free and Open-Source Framework—Python-Based Object-Oriented Programming for Gas- and Steam Turbine Cycles. *Energies* **2018**, *11*, 2609. [CrossRef]

15. Vincent, P.; Witherden, F.D.; Farrington, A.M.; Ntemos, G.; Vermeire, B.C.; Park, J.S.; Iyer, A.S. PyFR: Next-Generation High-Order Computational Fluid Dynamics on Many-Core Hardware (Invited). In Proceedings of the 22nd AIAA Computational Fluid Dynamics Conference, Dallas, TX, USA, 22–26 June 2015. [CrossRef]
16. Richardson, I.; Thomson, M. Integrated Simulation of Photovoltaic Micro-Generation and Domestic Electricity Demand: A One-Minute Resolution Open-Source Model. *Proc. Inst. Mech. Eng. Part A J. Power Energy* **2013**. [CrossRef]
17. Lo Brano, V.; Orioli, A.; Ciulla, G.; Di Gangi, A. An Improved Five-Parameter Model for Photovoltaic Modules. *Sol. Energy Mater. Sol. Cells* **2010**, *94*, 1358–1370. [CrossRef]
18. Ahmadi, A.; Robinson, P.H.; Elizondo, F.; Chilibroste, P. Implementation of CTR Dairy Model Using the Visual Basic for Application Language of Microsoft Excel. *Int. J. Agric. Environ. Inf. Syst.* **2018**, *9*, 74–86. [CrossRef]
19. Oxley, R.L.; Mays, L.W. Optimization-Simulation Model for Detention Basin System Design. *Water Resour. Manag.* **2014**, *28*, 1157–1171. [CrossRef]
20. Feingold, A. Radiant-Interchange Configuration Factors Between Various Selected Plane Surfaces. *Proc. R. Soc. Lond. A Math. Phys. Eng. Sci.* **1966**, *292*, 51–60.
21. Krysl, P. *Finite Element Modelling with Abaqus and Python for Thermal and Stress Analysis*, 1st ed.; Pressure Cooker Press: San Diego, CA, USA, 2017; pp. 189–198.
22. Lam, S.K.; Pitrou, A.; Seibert, S. Numba: A LLVM-Based Python JIT Compiler. In Proceedings of the Second Workshop on the LLVM Compiler Infrastructure in HPC-LLVM'15, Austin, TX, USA, 15 November 2015. [CrossRef]
23. Alexander, M.; Kusleika, D. *Excel® 2019 Power Programming with VBA*; Wiley: Hoboken, NJ, USA, 2019. [CrossRef]
24. Parlier, G.; Guéguen, H.; Hu, F. Smart Brute-Force Approach for Distribution Feeder Reconfiguration Problem. *Electr. Power Syst. Res.* **2019**, *174*. [CrossRef]
25. Mangan, T. *White paper on Perceived Performance–Tuning a System for What Really Matters;* T Murgent Technologies: Canton, MA, USA, 2003.
26. Lilja, D.J. *Measuring Computer Performance-A Practitioners Guide*, 1st ed.; Cambridge University Press: Cambridge, UK, 2008.

MDPI
St. Alban-Anlage 66
4052 Basel
Switzerland
Tel. +41 61 683 77 34
Fax +41 61 302 89 18
www.mdpi.com

Energies Editorial Office
E-mail: energies@mdpi.com
www.mdpi.com/journal/energies

www.ingramcontent.com/pod-product-compliance
Lightning Source LLC
LaVergne TN
LVHW070637170726
843515LV00004B/269
* 9 7 8 3 0 3 9 4 3 3 2 0 9 *